WHO SHOULD READ THIS?

95% of the wine brands that are developed and launched each year ultimately fail. Just visit your local wine shop and you will see the same wine brands on the shelf year after year as the new brands quickly disappear, with only a few making it longer.

Since the main subject and purpose of this book are made very clear by the title, there is a good chance that just opening it establishes you as someone who can benefit from the deeply experienced information contained in it. But just in case you're still on the fence, I will explain further.

It makes perfect sense to assume that of the 60,000 wine brands currently in the marketplace, the wines with the highest ratings, the finest quality, or the coolest labels represent the top sellers. Many of these brands also come from some of the most architecturally spectacular wineries and are displayed in those wine cases you see stacked high like fortresses at the front door of your wine shop, which also gives the impression that these wine companies sell wine at a very fast rate. However, nothing could be further from the truth.

Great wine is not enough.

I've seen countless efforts and dreams to succeed in the wine market fall apart at considerable financial and emotional cost to the entrepreneurs. You may be one of the many who have tried and failed. Or maybe you're currently an owner or part of the management team of a wine company with the goal of rebooting an existing but struggling wine brand. Or, more commonly, you might be a newbie with the next great idea floating around in your melon, eagerly ready to embark upon a new wine venture.

All of these approaches are different, but they are also very similar in terms of passion, determination and the goal of ultimately making great wine and selling it. This is something I can very much relate to. You see, this has been the purpose of my career - developing new wineries and new wine brands or managing existing wineries that are owned by individuals or companies, all trying to improve, grow into the future, and sell wine!

I believe this book could be your greatest - and most likely least expensive - investment in any of these scenarios. I'm here to move sales and marketing to the top of your mind, or at least on par with the more immediate thoughts of winemaking, label design and winery. Most would think this is a no-brainer. Sales and marketing; makes sense. However, I have witnessed during my seventeen years in the catbird seat of this excellent industry that this is definitely not the case, and not just sometimes but in almost all cases! Too often, owners and managers don't recognize the importance of sales and marketing, finding it less interesting than other aspects of the wine business, sort of like that huge homework paper you kept meaning to get to in college but never really did until the day it was due.

I can comfortably stake my career on the simple fact that sales and marketing is as important to the survival and success of your business as making great wine is, if not more so. But please don't get me wrong - this guide will not try to convince you to lessen the importance of making excellent wine. However, it will encourage you to consider some important decisions that need to be made, and to prioritize those decisions from the very beginning, not waiting until it's too late as so many do. It will answer questions and impart philosophies and tactics that can save you precious years and countless dollars, not to mention heartache. It is here to guide you toward not just crafting great wine, but more importantly, having an ever-increasing number of wine lovers adopt your brand, taste your efforts, and experience your awesome winery. Ultimately, you will realize the dream of being an integral player in the wine industry as either an owner or contributor. I can assure you, it's a magical journey when done well.

WHAT OTHERS ARE SAYING ABOUT THIS BOOK

"Must read. Eric has bottled his vast experience in the industry and passion for wine into a clear and thorough guide to create and market a successful wine brand. Enjoy without moderation!"
Alain Barbet - former Chairman and CEO
Pernod-Ricard Americas

"Having personally been in the thick of so many new brands, brand extensions as well as a plethora of brand redesigns over the past four decades, if I had a hat on, I would be taking it off as a sign of honor and respect to Eric for this fine piece of writing excellence. It should be the Bible, if not a standard read as a teaching tool, for all in the marketing and sales world of wine and spirits.

I was captivated the moment I started reading this masterpiece as it is written in a fashion that prevents you from putting it down. Then again, when you do put it down, it is just to take a note, mark a page and pour another glass of wine!

Kudos to Eric for a job well done. I'm sure it will be an eye-opener for many, and a confirmation for others."
Randy Ullom - SVP, Winemaster
Kendall Jackson, Jackson Family Wines

"As a grower and winemaker, the challenges of production are hard enough. However, being a winery owner brings even more challenges, the most significant being the marketing and selling of what we have made. Eric brings years of experience to this puzzle and gives an insider perspective on the details of wine sales and marketing."
Nick Goldschmidt - Winemaker & Owner
Goldschmidt Vineyards

"Eric knows the wine industry inside and out, from marketing through sales. His knowledge base is extraordinary. If you're looking for a straight-forward approach to building a wine brand, this is, without any shadow of a doubt, the book for you."
Tracey Nauright – Wine Industry Graphic Designer
Round Like A Circle Graphic Design

"Eric Guerra's book breaks down every essential detail of building a great wine brand or successful winery."
Daryl Groom – Winemaker, Groom Wines / Colby Red / DRG Wines
Former Winemaker & Senior Red Winemaker for Penfolds Wines, Australia
Chief Judge and Partner - Press Democrat North Coast Wine Challenge

When Great Wine Is Not Enough

DISCLAIMERS

.

When Great Wine Is Not Enough

**A Wine Sales & Marketing
Guide For Wineries
Négociants &
Wine Brand Owners**

Eric Guerra

Reserve Tastings Wine Company LLC

COPYRIGHT

When Great Wine Is Not Enough
© Copyright Pending

Cover Designed By:
Eric Guerra

Edited By:
Mindi Abair

Proofread By:
Mark Rickerby
using www.fivwrr.com

Published By:
Reserve Tastings Wine Company LLC

Visit us at:
www.reservetastings.com
Instagram: @reservetastings
Facebook: www.facebook.com/reservetastings

Printed In The United States of America

DEDICATIONS

For Mindi, Jordyn and Madison

Il Mio Amore, Il Mio Cuore, Il Mio Animo
My Love, My Heart, My Soul

Also, for all the wine professionals who endure endless hours each day passionately selling and marketing wine with the purpose of putting a glass in every wine lover's hand. Cheers to you!

TABLE OF CONTENTS

PREFACE

"A bottle of wine contains more
philosophy than all the books
in the world."
- Louis Pastor

As I write this, I am only seventeen years into a career and journey that has been largely . . . *unexpected.*

It is best to understand upfront that I'm not a scholar writing a book based on tons of data about the wine business. The information in this book is derived from my direct experiences and interpretations of what worked and what didn't work when I managed some of the greatest wineries and wine brands in the world. I've been fortunate enough to meet and work with the most accomplished people with grand dreams of creating the next legendary winery, as well as winemaking geniuses as they crafted some of the most incredible wines the world has ever experienced.

I've walked cool, damp miles in chalk caves beneath the luxurious Avenue de Champagne in Épernay. I've marveled at the sunrise as it cast a shadow over the tiny Grand Cru vineyard of Les Clos on the edge of Chablis. I've sipped epic Petite Sirah poured from mini reused vials in a small, funky tasting room with a winemaker who had more wine stains on his shirt than there was wine in my glass. And in pure joy late in the workday, I sat alone on the terrace of a renowned boutique winery listening to the absolutely impeccable voice of Donny Hathaway singing *A Song For You* as it resonated high above the trees in the Napa Valley.

During each of these moments, I thought to myself, *"This has to be right. It's just too perfect."*

All the artistry, the hard work, the passionate and dedicated people, the breathtaking views, the history, the heartbreaks, the rewards, and the unbelievable process that is required to make a simple bottle of wine appear on a table is truly remarkable, let alone millions of tables for millions of people to enjoy every night.

As a contributor of these wineries and wine brands, I've seen many succeed and others fall far short of expectations. As a leader of other wineries, I've experienced the same ups and downs even more deeply, wondering later what I did right and what I could have done better, the latter usually dominating my thoughts more. Having this perspective constantly reminds me that I am a mere mortal, a wine-geek with still so much to learn. It also comforts me because I know the hard lessons will make the next venture much more successful. This is an industry of endless learning. As in all things, the key is to never make the same mistake twice.

This remarkable industry has given - and taken - so much from me over the years. And as I've become more experienced, humbled and knowledgeable of this complex business of wine, I've always been interested in those who come to these places of pure beauty with ambitions beyond just wine tasting. I watch as they are drawn in by the Disney-like experience we create for our customers. As they are leaving wine country, I often hear them say to their significant others, with the blind yet bold confidence, usually developed by success in some other industry, that they will be back someday to fulfill their life-long dream of creating a storybook winery and the next legendary wine brand. Of those who follow through with this dream, most inevitably spend countless millions then fail by making the same mistakes - mistakes that stem from a simple lack of knowledge and experience. I know this because I did it, too, in my early years.

I have also seen countless small wineries owned by unwitting grape growers, or entrepreneur négociants with the next can't-lose brand idea, or even mid-sized wineries with an experienced and professional wine team, all make many of the same mistakes and add their names to the brands that get tossed onto the "discontinued" pile every year. All those dreams reduced to dusty cases of unwanted wine piling up in warehouses, only to be written off as a huge financial loss. The psychological toll is sometimes even more damaging as they ponder the countless hours spent creating an excellent wine brand only to watch it march slowly toward its final resting place as a full bottle of acetic acid, more commonly known as vinegar.

What I have learned is simply that ***great wine is not enough***.

Almost everyone who takes on the challenge of making the finest wine or greatest brand in the world starts with the purest sense of purpose and the most passionate intentions. Wine is a luxury to most and being part of this artisan process is a highly appealing endeavor to achieve in one's life.

This is the very reason I decided to write this guide - if only to help one person, one winery, one négociant, or one great brand idea get to that place of accomplishment so they can know that their time, money and dream were all worth it in the end.

I believe the craft of winemaking is the greatest passion and highest form of artistry there is. And if you have taken on this challenge, I don't want to deter you. Actually, I want to encourage and commend you for taking the leap, as I did many years ago. It is the greatest industry in the world. So, congratulate yourself for taking the first step. Then, after all the handshaking and self-adulation, I strongly advise you to hold on tight because what you think will happen and what will actually happen will be vastly different, I promise you. Even with my Malcolm Gladwell-required 10,000 hours of wine experience under my belt, there isn't a day that goes by that I don't experience something new or have something come up that was completely unexpected. I've been told that others can see the confusion, that always precedes growth, on my face from way across a conference room. Apparently, it manifests itself by me rubbing my forehead, accompanied by a perplexed, almost clueless look while I'm trying to figure out something I never experienced before. The more I think about it, it is much like something my New Jersey-Italian father did at the dinner table when I was in high school circa early 1980's and he found out about something I shouldn't have done. He would cover half of his face with one hand, then firmly slide it down as if wiping away the disbelief, until his hand and he finally arrived at a place of semi-acceptance. I finally get it. *Sorry, Dad.*

Now that you have this book, your first step in creating a new winery or new wine brand or redoing a poorly performing wine brand is simple. For the newbies, however, it does not start by what you think should happen. Please do not listen to the voice in your head. You know, the one that says, *"You got this. You know this. You've done this before."* Or, the more dangerous voice that says, *"Get out the pen and start writing open checks for wine, winemakers, architects and label designers."* Step away from the pen and promise yourself you will learn from others, starting with this book, and broaden your perspective by placing an equal focus and investment on all parts of the business. Put sales and marketing first or at least equal to the shinier, sexier objects like wine, the label and the winery. If you do this, your thinking will already be vastly superior, and you'll have a huge advantage over more than 95% who came before you and failed. I use that percentage with great accuracy. I'm not trying to be sensational. (Well, maybe just a little, but you get the point. It's a lot.)

I am confident that the thought process and proven steps contained in this book will reward you as much as learning them firsthand has done for me. I hope you enjoy this guide as much as I enjoyed writing it.

So, grab a goblet of wine… I'll join you and do the same… because here we go!

Cheers!

- Eric

SECTION I: THE PASSION STATEMENT

"[I]t is the wine that leads me on,
the wild wine
that sets the wisest man to sing
at the top of his lungs,
laugh like a fool – it drives the
man to dancing... it even
tempts him to blurt out stories
better never told."
- Homer, The Odyssey

I'm going to assume you made it here because you are about to embark upon a new wine venture, or possibly a reboot of an existing wine brand. If this is the case, again, I congratulate you on overcoming the first great challenge, that of taking the initial, real step toward building your dream wine initiative. As I'll outline in more detail in the following sections, and possibly repeat more often than necessary because I'm pretty excited about it, I believe there is no greater journey to embark upon. The wine industry is magical, mystifying, awe-inspiring, romantic, artisanal, intriguing and cerebral. There is a reason why, every year, so many new brands are introduced into the marketplace and droves of people seek careers with all the different wine companies, and that reason is that it is simply just that damn awesome. Working in wine doesn't feel like a job; it is a lifestyle and a truly noble purpose.

There is no place more important to start than with this very simple, yet profound step . . . by developing the purpose and passion behind your wine venture. This doesn't mean to envision it and tell your family and friends. That's how 99% of the other wine ventures typically begin. It means thoughtfully thinking it through and writing down your statement. You've heard this exercise preached by the greatest motivators of our time like Steven Covey, Anthony Robbins and Steve Jobs, and with good reason. It is your True North Star when ramping up the business, developing the winery, creating the brand, figuring out the thousands of details, and pulling together the team. More importantly, as the daily grind kicks in and you are deep in the weeds - which is guaranteed to happen because it is a byproduct of our very complex industry - it is a statement to remind you what was in your mind and heart the day you embarked upon this special journey.

Without it, at some point your business will be at significant risk of running off the tracks because it will not have the founding purpose and defined direction so necessary to survive, let alone succeed. It is all too common for the original purpose to gradually become lost and forgotten as the litany of details and major challenges become daily activities. A great majority of wine companies I've been a part of have become something very different than what the owners or founding team set out to accomplish. On the other hand, I have also managed a number of other wine companies that are highly successful and some of the most legendary brands in the industry. And while working there, it was easy to see why - because the original purpose and passion of the organization – the mission – was frequently reinforced by those at the top.

Here is a short story that will convey this point, and something I think you will very much appreciate.

A Profound Message from the Wine Legends

Many years ago, I had the opportunity to spend time with Robert and Peter Mondavi during an auction event. At the time, I was managing the iconic Napa Valley, Rutherford-based sparkling wine brand called Mumm Napa. I was told that the Mondavi brothers were friends of the late winemaking and founding member of Mumm Napa, Guy Devaux. So, I thought we had a little connection I could use to start a conversation. What happened next was we were positioned together while waiting for the next schedule of events of the auction to begin. I knew I had a unique moment in time to ask a few questions and learn as much as I could as quickly as possible. I found the moment, broke the ice, and it developed into just the three of us freely chatting about the industry. At this time Robert was in his early 90's, mostly confined to a wheelchair and not the gregarious, outspoken man who became a legend in the wine business decades earlier. However, he was aware and easily following the conversation, occasionally chiming in. Peter was very conversational and extremely mindful, still heading up Charles Krug Winery, which was the original Mondavi family winery created by their father, Cesare Mondavi. In the world of wine, especially luxury wine, there are few people on the planet that are as renowned, entrepreneurial in spirit, and experienced as these two.

We discussed the state of the wine business, the evolution of Napa Valley, the happenings at the auction and a few other random wine topics. Candidly, I was trying to listen, but my mind was swimming with the fact that I had this moment. I stayed focused as much as possible. As we chatted about wine and the auction, I realized we only had a few minutes left. I turned to Bob and asked him a very specific question, one I thought might provide an answer I could take with me for the rest of my life in wine.

I asked, "Mr. Mondavi, still being early on in my career, I was wondering if you could give me some advice about how to be truly successful long-term?"

There it was, the million-dollar question thrown out there into the air. Almost like asking Buddha what the meaning of life is - in wine speak, of course. Remember, at this time Bob wasn't his typically extroverted self. His answers up to then were a little unhurried, and Peter, in his distinguished broken English with a generous amount of Italian sprinkled in, did most of the talking throughout the conversation. What caught me off guard was that *this* question, for a fraction of a moment, brought excitement to Bob. I'll never forget it. All the people and sounds surrounding us disappeared and the moment seemed to play out in slow motion.

His eyes brightened up and focused very intently on me. He paused, I believe to gather his thoughts, then said, "Eric, as you get higher up and you become very busy dealing with the challenges this business brings us all, never forget the reason why you got into wine to start with. Remind yourself often. Never forget." Then he slowly and confidently smiled. I can see Peter out of the corner of my eye nodding in agreeance.

It has been well-documented in great detail that these two wine brothers did not get along for many years. Additionally, years ago, Bob took his wine company public, being one of the very first to try it. I believe he experienced how it changed his winery dramatically, twisted some of the passionate reasons he started it, and negatively influenced the decisions he was then forced to make. In some profound way, I believe he was speaking not only from his heart, but from a vast amount of experience that few have.

Now fast-forward to this moment – they are together in the twilight of their lives, seemingly getting along in the spirit that *the past is the past*, both agreeing about a profound notion, and teaching it to a wine industry greenhorn - that no matter what happens, even very public family conflicts and burdening financial obligations, there was a day when we all sat alone with our thoughts and made the decision to be in the wine business, with the singular passion that we just love wine, and to never forget this very simple fact.

And with that, Bob continued to smile at me as his nephew grabbed the handles of his wheelchair and started to push him away to the next event he was scheduled to attend. As they left, I didn't say anything back to him, as his words were so simple yet so profoundly perfect. I know he knew I got it as we smiled at each other. It was the perfect answer. The meaning of wine-life offered directly to me in this special moment in time.

Never forget the reason why you got into wine to start with. Remind yourself often. Never forget.

All these years later, his words are still with me, as powerful as they were when he said them. However, I really didn't anticipate how this simple statement would come back to me over and over again and guide me through some very important decisions in all aspects of the wine business, especially at the toughest moments.

Most of us get into the wine business for the pure love, passion and joy it brings to us, as well as millions of others who drink our wines every day. As an executive, I've had to make some decisions I wasn't necessarily proud of. Some decisions had to lean toward the bottom line rather than the best interest of the craft. But overall, when it mattered, when the day would get crazy with details and I was unsure about the future of a winery or wine brand, this simple statement served me well. I have it in my iPad as an every-day reminder, and you should as well.

It has never let me down.

Start With Something So Simple, Yet So Powerful

It is time for you to put pen to paper, or keyboard to electronic notes. Your first step is to begin by clearly defining why you want to build your dream and put all your passion into wine. Think back to the day it hit you. It could have been while driving through wine country like so many other dreamers, or it could have hit you while sitting at home drinking wine and becoming more intrigued by each glass that blew your mind. Whatever your *Aha!* moment was, go back to it and recall the feeling, the emotion, the interest, the intellect, the intrigue, and the passion it stirred in you.

This is pretty much a mission statement, but since we're dealing with wine, I call it "The Passion Statement" because it's why just about all of us do what we do every day, for the passion and love of wine. This statement can be whatever you want it to be, detailed or simple, but it needs to come from your right brain, the emotionally creative side. Make it an imaginative, emotionally raw, dream-laden, holistically-thinking declaration. It could be about crafting a superior wine, or living the luxury wine lifestyle, or managing a winery at the highest level, or having a winery that people visit from around the world. Write it down and keep it in a place you can recall frequently.

This Passion Statement is your guiding light, your True North Star, and your advisor in times of decision-making. And if you're bold enough to open the kimono, it's something you should share with your team. Everyone should be on the same page. Go as far as posting it in the winery back office. It will have a profound effect on you and your team, not only as a reminder of why we deal with the daily difficulties of such a challenging business, but as a benchmark of the expectations of excellence you require. Most importantly, use it as a barometer for making decisions when you're in the weeds or when the purpose you started this journey with is not at the top of your mind for whatever reason. It will always guide you to the right decision and the right place, regardless of the situation at hand. And for others to follow it, you need to embody it in everything you do, or it is meaningless. Walk the walk and others will be proud to follow in your footsteps.

Finally, The Passion Statement is something that is set in stone and lives with the winery and brand in perpetuity. When completed and posted, it is no longer yours, it is the wine venture. Imagine if on the wall of a 200-year-old Bordeaux winery there was a neatly framed statement from the original owner stating why he or she started the property, why they loved the region, why they got into wine, and expressed their overall purpose and passion. Think how powerful that message harkening from the past would be to the people making decisions today. It could be the most important, yet simplest, task you take on at this very moment.

Your Passion Statement

_____

_____

_____

_____

_____

_____

_____

_____

_____

_____

SECTION II: PUTTING SALES & MARKETING FIRST

"Making great wine is very difficult....
but selling it is even harder."
- Quoted by almost every wine veteran

If this is the first time you've heard this well-known quote, don't take it the wrong way. It is not intended to minimize the enormous effort and skill that goes into crafting fine wines. It's more about how the sales and marketing aspects of the wine business are often overlooked, or how they are considered well after the wine and winery plans are completed, and many times even after the wine has been released to the public.

I hope you've been fortunate to weave along the Silverado Trail in Napa Valley, or the rustic West Side Road of the Russian River Valley Sonoma, or the prestige Avenue De Champagne of Champagne France, or even the legendary Strada Provinciale Bolgherese of Bolgheri, Italy. If so, you have gazed upon some of the world's most brilliantly designed wineries, neatly placed within the vineyards, slightly hidden behind well-manicured hedges or picture-framed in the distance by a monogrammed iron gates. Each has their own story of special family histories, new beginnings, retirement rewards from a previously successful career, or an established wine company portfolio investment. And, typically, they are owned or managed by incredibly capable, passionate people who believe that the dream of making a luxury wine is a conviction that can be achieved and lived every day.

Or, more commonly, you have probably strolled down the many aisles of your local wine retailer and gazed at the thousands and thousands of wine labels, each looking perfectly designed and outwardly having their own story to offer. And since there are so many bottles on these shelves, it would be logical to think how easy it must be to own a brand, put wine into a bottle, create an artsy label and away you go to happy wine-sales town, with dreams of masses clamoring to buy and drink the wine.

Many years ago, I was one of those idealists. They say when you find your calling in life, it grips and takes hold of your every thought, giving you great purpose and meaning. This happened to me in the very same way it has done to so many others with those wineries you see behind the gates or those brands displayed prominently on the retail shelves. In this order - my family, wine and Italian food took over my every thought and became my daily determination. Diving into the endless details about winemaking, grapes, regions, vineyards, barrels, wine styles and label art - while eating pasta, of course - was endlessly fascinating to me. To this day, I study everything I can get my hands on that relates to wine. I just had to be a part of the wine industry in some way because it was just too engaging and perfect to be a casual hobby. So, I did what so many do every day. I cleaned my rosé-colored glasses and walked into the deep end of the industry. I left my career in high-tech and embarked upon a journey into what I saw as the Disney-like world and lifestyle of wine. As with most luxury industries, what I quickly experienced in the day-to-day grind was a bit different than what I envisioned, and very different than what I was marketing to the wine consumer.

I immediately learned that the picture-perfect experience was largely an appearance of what the industry wanted you to see . . . the sexy, awe-inspiring experience showcasing a luxury lifestyle, fine art and killer wine flowing freely about with each brand having its own adventurous story. And for the financially fortunate within one of those architecturally impressive wineries, it is pretty awesome. Even though I now have many years of experience, having seen countless such wineries, knowing their inner workings and how they put on this veneer still inspires me to this day. It's the reason so many try to begin a wine lifestyle in this crazy industry - seeing the payoff when thousands experience your wine is truly excellent. Regardless of what goes on in the background, we all want to believe in the dream, and when it is presented in such a spectacular way, we buy into it every time regardless of our knowledge or understandings of the non-sexy, tough parts.

With that said, as you may already know if you are a tested wine veteran, or if you're just starting to experience the craziness as a newcomer, the tough parts are very tough. Without overstating it, 97%

of the time is involved with the behind-the-scenes grind. There are challenges consistently with rising costs, a poor growing year, federal and state government compliance hurdles, vendor errors, winemaking difficulties, aging inventory, problematic distribution, costly fulfillment, and ratings not at the level expected. This list may seem long, but conservatively, it is a quarter of what typically is dealt with in just one day. I tell you this not to deter you from the industry, but to make you aware of the level of complexity and never-ending unexpected challenges that will come at you once the brand is up and running. The wine business is not for the weak of heart or soul, and especially not for those that cannot financially handle the ups and downs.

Many will take on the climb with the enthusiastic expectation of reaching the summit, and they will typically start at base camp with the wine and winery driven approach.

The Wine & Winery Driven Approach

Each year, countless droves of happy people visit our little slice of heaven called wine country. They fall in love with the incredible allure, sexiness and luxury lifestyles we enjoy here, then seek to live the ultimate wine life they see before them. They unload a sizable chunk of their bank accounts, move to the greatest place in the world, build a storybook winery, and spend years crafting a world-class wine. Or, on the other hand, there are those instances where they may not have the financial assistance from previous career successes or the backing of a wine company's treasure chest. They enter into the wine business out of pure passion and love for the wine, calling in favors, sleeping on friends' couches, and using small bins tucked away in in the back of the wineries to put their dreams together. And if they have the patience, and the near-endless stream of money or favors from friends, given many years of trial and error, most of these dreamers can get pretty far making a very fine wine.

Then, as the wine gets closer to the exciting and much anticipated release date, reality hits as they quickly recognize they did not give the same level of thought and planning about how to sell the wine that they did to crafting it. Regardless of how excellent the winery, the wine brand and the label are, this is where most falter. The unexpected daily grind takes the majority of new wine ventures to their knees because they didn't put a quality sales and marketing plan in place, or the right people to implement it, *at the beginning of the process.* As they build their brand or winery, they focus the majority of their time on crafting the wine, winery and label. Then, once these pieces are in place, the reality starts to sink in that they need to sell all the pallets of wine they just spent huge amounts of money creating, so they turn to sales. What comes next is a sales and marketing plan that is usually self-developed, using a fraction of the expertise, investment and thoughtfulness that was spent making the wine. Most of these wine ventures are not profitable, or profitable enough to make them sustainable in the long-term.

Without exception, every new wine entrepreneur starts with hiring the winemaker, a label design firm, and a local architectural firm if they are building a winery. However, I couldn't tell you of a single time when I've heard a sales and marketing wine professional was hired at the same time or even before the winemaker. It's a very fascinating and unfortunate process. Just as there are trained winemakers, there are as many extremely experienced and talented sales and marketing professionals in the wine industry, and for great reasons - the most important being that it is difficult to make great wine, but selling it is even harder.

I understand why sales and marketing can be overlooked so quickly and easily. The excitement of the wine and label designs are the first experience many newcomers have exposure to well before they are in the industry. Crafting the wine and the label seem so much more appealing, artsy and interesting. And candidly, it is so much so that insiders unwittingly get caught up in it as well. But when the newcomer visits a winery during the dream-phase, what is not presented along with the Disney-like experience is the enormous mountains that need to be climbed by the sales and marketing team every hour of every day. Interestingly, when some now mature wineries started out, they had 70% of their employees on the winemaking and production side, only to have that shift to 70% of their employees on the sales and marketing side due to the pure need for survival, let alone prosperity. We all learn the immense importance of wine sales at some point or another, but usually too late for most.

The Sales & Marketing Driven Approach

Those of us who are experienced in the wine business and fortunate enough to work for wineries as our careers get to see it over and over again. Logic strongly suggests that the best wine brands with the highest quality wine should be the industry leaders, but many times nothing could be further from the truth. With over 60,000 wine brands in the US at any given time, with technology and winemaking expertise at incredibly high levels, with an ocean of truly outstanding wines in the marketplace, with countless stunning wine labels, and with even more wineries being recognized as architectural marvels, survival comes down to those that have well-thought-out and executed sales and marketing plans *every time*. The industry has numerous examples demonstrating that the wine companies with the most focused sales and marketing strategies, along with motivated, expert teams that have bought into and embody the vision, are now and will always be the industry survivors and leaders.

Great wine is not enough.

This is the wine-law engrained in the soil for years. Robert Mondavi recognized and then revolutionized it. Unfortunately, most others have not, and bankrupt wineries are easy to see as you drive through wine country because the names on the signs outside their gates change from year to year. Yes, of course the wine has to over-deliver and be superior. Don't get the perception that this guide will minimize that part of the business. It's a given that a wine venture needs to commit to having the best wine they can create, without question. However, there is one simple fact that cannot be disputed . . . great wine does not get tasted if no one buys it.

To succeed with a new wine venture or a rebirth of a struggling one, it is imperative that you place sales and marketing activities and outstanding, experienced personnel on the same level of importance as the investment you are willing to make to source or create the wine and winery. And this certainly does not mean *after* the launch of the winery, but at the inception.

Let's dive a little deeper using a few common scenarios. Which one are you?

Scenario 1 – The All In

If you're going big and starting up your own winery by either building it from the ground up or buying into an existing property that you will reestablish under your brand, the effort and time of pulling all the winemaking pieces together are extensive, with massive expenses. By now you should be very familiar with my thoughts and concerns regarding outsiders entering into the wine business, spending big dollars on physical wineries only to struggle or go under due to a lack of sales and marketing sophistication. With that understanding, I will safely assume that you are in it to win and willing to take the more well-informed path. Therefore, since you're eager to take the bull by the horns, if you do this first step I'm about to outline, your likelihood of success will go up exponentially without a doubt.

It is imperative that you put first things first and immediately start a search to hire a very experienced sales and marketing wine professional. This person can not only start building the sales and marketing plan while you're creating the winery and wine program, but also take on the multitude of other non-winemaking initiatives that the business will require. There are many highly capable and tremendously intelligent wine professionals who can come onboard and wear numerous hats because their years of experience in other wine business environments has taught them more than just one trick. But here are a few things to keep in mind when searching for the right person.

Make sure they have a strong background in sales or marketing that is in wine, preferably in wine companies similar to yours. I would advise not to entertain anyone from outside of the industry at this level, regardless of how capable and intelligent they may seem, because it is too much of a risk. Your business is not the right training ground for a new career path. Those newbie positions are best left with the largest wine companies that have the programs and infrastructure to nurture and give time to teach the basics. The wine industry has many intricate details that are very specific to wine and that can only be acquired by direct experience while in the trenches. This is a non-negotiable fact. In later sections, to further assist you in finding the ideal sales and marketing people for your team, we will go into detail regarding wine sales representative personalities and types of sales roles.

After you find your rock star, you will most likely be giving them their marching orders. When doing this, keep in mind and make sure to have them focus a majority of their time on sales and marketing. I know this sounds simple, but I'm speaking from years of experience here. When the madness of our industry starts to seep into their daily activities, they can easily get caught up in the litany of other fire drills that go on during a normal wine day. Therefore, it is truly imperative to make sure they have very clear direction from you that no matter what happens in a day, they need to focus on the big strategic tasks and not always the little reactive tasks. Finally, professionals like this are not inexpensive. It is a one-to-one ratio in terms of value-to-quality of work and experience. Think of it as an insurance policy for what you've already invested in terms of dollars, time and resources. In the end, their high compensation should be well in proportion to the overall gain of a successful wine venture.

Scenario 2 – The Brand Builder
The next scenario is if you are starting your own wine venture, but as a négociant where you will be creating a brand but not a winery. If you're not very familiar with this method, it happens more often than not because it is generally the lowest cost method to get into the wine business, the main reason being you don't have to spend the millions of dollars required to build or buy a physical winery. However, even though it does not require the extensive capital on the winemaking and winery side, it does require an even greater investment in sales and marketing than most other wine business models. When owning a winery that has full visitation access, or possibly a more limited by-

appointment-only hospitality experience, by default, these companies are exposed to a stream of visitors they can sell wine to with little effort. Many factors go into how much they can sell, such as visitation restrictions, location, accessibility, and their capability to host guests on the property. However, wine lovers love to visit wineries, even those that are located at the far reaches of a wine region or those that are not very visitor-friendly. I know that may be a bit shocking, but I've visited many wineries that can sell some wine in spite of their own terrible customer service. But, as mentioned previously, if they don't have a well thought-out and well-implemented sales and marketing plan, they too will find it difficult to survive, and most do.

On the other hand, starting as a négociant and building a wine brand from the ground up requires an even more focused and perfectly executed sales and marketing plan to make up for the lack of visitor experience and low-hanging sales some wineries take mostly for granted.

As one of my many career stops, I was part of a winery that sold wine to hundreds of négociants in the US each year, from existing, mature wine companies to start-ups, which was the most typical customer. What I saw nearly 9 out of 10 times was disheartening, to say the least. Without exception, they all had the next great brand idea and label design thought out, the easiest and most common items to tackle for those with no history or experience in the wine industry. They are the shiny, fun objects that usually get most of the attention. They were all overtly confident that they knew how and where to sell the wine, which I found interesting because the largest and most sophisticated wine companies in the world struggle with this topic each and every day. But these outsiders, usually as a result of having some level of success with a completely unrelated business, had figured it out. And because they had this formula for success and such bumptious confidence that they would succeed, almost all did not hire an experienced sales and marketing team. If they had anyone part of their unexperienced team, they usually were someone from a prior, non-related wine business or, worse, a free family member. They would sign contracts for tens or sometimes hundreds of thousands of dollars for wine and wine services, with little to no wine experience on the team and with nothing more than a good idea and a nice label design.

Now, I don't want to come across as an old wine guy and suggest the wine industry does not welcome entrepreneurial approaches and people. If anything, I am more of an advocate of the thought process that doing things differently is always better than the fifty years of the same techniques and ideas. There's definitely a place for newness and innovation in the industry. However, given its sophistication, very specific requirements, legalities, and just general ways of working that are engrained in the "system" of selling alcohol, it does require some prior level of sales and marketing experience and expertise, and the more the better. Yes, a very few completely inexperienced négociants have reached some level of success, but that is very much the exception rather than the rule. However, the amount of capital and time wasted getting there is usually so significant, only a few can afford to finance creating and implementing a business model that requires years of catching up just to squeeze out a small profit, if any.

So, if you have a great wine brand concept and are going down the négociant path with a big smile on your face, above all else, your first step is to invest in and hire two experienced wine professionals well before securing the wine. One salesperson and one marketing person. The importance of sales and marketing to your companies' success is proportionately large and the amount of details needed for the development and execution of the strategy, as well as the daily grind, is significant for both roles. It absolutely requires two dedicated, excellent people.

I'm aware that many négociants are not located in a wine country but are typically in larger cities around the country. If this is your situation, this does not give you a hall pass to hire a distant relative, or a favorite uncle you owe a favor to, or any other inexperienced wine person. Though there might not be as many deeply-skilled candidates in the city as you would find in and around wine country, it is a huge, multi-billion dollar, national industry today, so finding two rock stars is a strong possibility if you make it a priority and treat it as important as finding the wine or developing the brand.

I'm guessing right about now you are hearing this for the first time and questioning the notion of hiring two wine people well before obtaining the wine or designing a brand. I bet in your mind you are negotiating with yourself that you can get by with just one person. Or you are possibly justifying using your significant other or a relative who might be in place already. Remember, I've come across hundreds of these scenarios, so I'm one step ahead of you and here to be the voice of experience and reason, as well as to remind you that neither is a possibility. Just know that you will be compromising the very idea, the large investment and - let's be very candid here - the passionate intent behind your wine company. Just having a good brand concept and great wine is not enough. Beginning the whole process with a quality, strong team experienced in wine sales and marketing will put you years, and hundreds of thousands of dollars, ahead of everyone else. At this very moment, believe and embody this ideal and you will have a much better chance of experiencing great success.

Scenario 3 – The "Never Say Die"

The final scenario is if you are part of an existing wine company and looking to launch a new wine brand or overhaul an existing one. This is a very common situation because it is extremely challenging to build a successful wine brand and manage it for many years at that high mark. Few are able to reach this level. So at some point, the owners of these declining brands either make the decision to cut their losses and discontinue the brand, or they believe there is some consumer equity remaining in the brand name to give it another go. Only you can make that decision because it is largely an intuitive one. If you believe the brand has some life and love left in the marketplace, you might owe it to yourself to give it another chance. There are numerous well-known brands that have declined and come back to experience some level of success, but I'm afraid there are even more examples of those that don't make it back from the long march toward their inescapable fate. So, if you have visited your therapist's office to discuss the finer emotional details and your decision is to give the brand a shot at a rebirth, then let's move forward with positive conviction that it can and will be done successfully.

Here I'm going to give you a little different advice than the two previous scenarios. Since this is an existing wine company, I will assume there is a solid sales and marketing team in place. If this assumption is not true, then rewind to step one above and meet me back here when you hire a few experienced sales and marketing people.

First, evaluate the wine. I know, this is a complete departure from the focus of this entire guide, but I've found that if the brand has some solid consumer awareness and equity, and you slowly lost customers along the way, then something intangible and not readily apparent made them bounce out of the brand and not stay long-term. That can only be one of three things - wine quality, price or presence. I don't believe label design impacts this decision on a large scale. I've seen both the simplest to the most peculiar brand designs sell fine. So it is probably that of the wine style chaning or quality tapering off. If so, it is important to take a step back and very honestly discuss it with your team, evaluate the problem and take corrective actions as the new brand is being developed.

Second, evaluate the price. Maybe the wine quality and style are on target, but the marketplace has thousands of competitive bottles sitting right next to your brand, or hundreds of other wineries and tasting rooms to visit every day. And maybe you made the decision to not keep up with the endless promotions and discounts due to financial pressures, or not to stay competitive with your hospitality tasting fees, thus giving your customers an easy out to explore other wineries and wines. There can be an entire book on the strategies and complexities written on this subject matter alone. However, later on in the guide there is a pricing section to assist you at a higher level. Wine and hospitality pricing are an extremely intricate and detailed process for most wineries and négociants, so make sure to cover this with your team and have a very solid understanding of what is broken - the brand concept or your pricing.

These are two very involved topics, I understand, but important for you to at least be aware of prior to getting to this concluding point.

If you have completed your analysis on wine style, wine quality, pricing and brand health, if you have your therapist's okay, and if you have determined that giving your brand a make-over is the best course of action, then I commend you for making it this far. It is more an emotional decision than an academic one because your brand may have lived a full life prior to this point, giving you the means to build a great wine company and provide for your family, and offered you the chance to have a career in this great industry. Letting go sometimes is not an option. So, let's shift into success-mode here and understand what it takes to move your brand forward.

Since you already have an established wine company at some level of success, you might have the financial stability to have great people in the sales and marketing department. However, though the first step in rebuilding is making sure you have the right team in place, this doesn't always mean the most capable or fastest. I believe there are two types of experienced wine sales and marketing people. The first are those who are über capable, can pull together complex data within an afternoon, build a full plan by the evening, and create a whole new strategy by the end of the week. These type-A's are usually very intelligent and motivated and go about their flight to the top of the ladder by following the MBA handbook of business practices. Everything they do comes from some level of business background, with sprinkles of experience thrown in.

Then there are those who are just as capable but are driven be years of experience and wine intuition. Of course, having a solid education, even at the MBA level, is always a bonus, and the importance of that should not be minimized. Nonetheless, I've seen many people assume that just having a high level of education must mean they can excel at building and executing a strategy based upon wine. I say this from my own observations of working for and with the supervisors of hundreds of wine professionals, and actually being one. Those that have the most intuition about the wine industry, which almost always comes from direct experience, are by far the ones who think the most comprehensively about the challenges ahead and have the greatest understanding of how to build a successful plan. The path forward for rebuilding a brand takes some level of prior experience that can only be obtained by being in similar situations in the past.

Evaluate the team for their strengths and weaknesses and have those with the most intuition and experience on this project. They may not work the fastest or have the shiniest presentations, but they have their finger on the pulse and will be the key people needed. From there, they will drive the brand strategy, wine style, packaging, and everything in between to set your brand up the best for another go.

Those that understand this early-on have a high success rate, or at least spend a lot less money getting to their goals. Just look to Robert Mondavi's legacy for the proof.

There Are Two Paths, So Take The One Less Traveled

There are many details and intricacies on the wine production and vineyard side of our business. Since it is much more appealing, there are countless books, websites and experts available to interest (or bore) you on any given night. I believe this guide may be one of the first that approaches the idea that sales and marketing needs to be at the same heightened level of awareness. It needs to have the same number of classes available in universities along with the Enology programs. It needs to have the same level of rock-star status and consulting fees as winemaking and vineyard management. It needs to have the same cultural exposure, expectations and perception of excellence in all wine companies. And it needs to have the same frequency of discussions between wine business associates at the Napa Valley Coffee Roasting Company on a Monday morning. Can you tell I am cut from the mold of wine sales and marketing? Ya think?

I've directly experienced how difficult it is to sell wine. It is a daily, at times soul-searching, endeavor. There is a very good reason that sales and marketing professionals have a higher compensation package than most others in the wine business. But if you develop this part of your venture too late, it is not something that can be caught up too quickly. All the investment in winery personnel, vineyards, wine, and production will rapidly weigh down the organization so much, it will sink the ship shortly after leaving the dock. Only those with deep pockets who can right the ship and withstand the challenges will make it through, and even then, most scuttle the ship at some point.

Here your path diverges to either focus on wine, winemaking and label design, or to focus on sales and marketing. Be one of the few and take the path less traveled, because it will make all the difference.

Two roads diverged in a yellow wood,
And sorry I could not travel both
And be one traveler, long I stood
And looked down one as far as I could
To where it bent in the undergrowth;

Then took the other, as just as fair,
And having perhaps the better claim,
Because it was grassy and wanted wear;
Though as for that the passing there
Had worn them really about the same,

And both that morning equally lay
In leaves no step had trodden black.
Oh, I kept the first for another day!
Yet knowing how way leads on to way,
I doubted if I should ever come back.

I shall be telling this with a sigh
Somewhere ages and ages hence:
Two roads diverged in a wood, and I—
I took the one less traveled by,
And that has made all the difference.

Robert Frost

SECTION III: WHAT YOU NEED TO KNOW TODAY FOR GREAT SUCCESS TOMORROW

"Nothing makes the future look so rosy
as to contemplate it through
a glass of Chambertin."
- Napoleon

This section came entirely from that inspiring conversation I had with Robert and Peter Mondavi I wrote about earlier. If that moment did not occur, I would have never considered covering this topic so early on in the process because the subject matter seemed so simple. If success comes in your wine business, you continue to grow and push the boundaries, right? It is a question that has to be discussed amongst all the wine brands and wineries that exceeded their original expectations. The customers love them, and they were clearly doing something right, so it just made sense to keep growing beyond their plans – to make it happen, take on the world, be the biggest and best.

However, the actual answer might surprise you, as it did me years ago. Maybe history has proven that in the wine business, success does not bring on more success . . . if it requires you to change your original vision and strategy.

To Scale or Not to Scale

Referring back to my conversation with Robert when he said to me, *"Never forget the reason why you got into wine to start with. Remind yourself often. Never forget."*

He was asking me to think back on the core reasons I got into the wine business when I was starting out. He suggested I should contemplate those purposes often, especially when the day gets busy and the business becomes more complex. But I also believe that statement came from a man who had to deal with making very impactful decisions during times of great success; decisions he possibly later regretted once he had a moment to think back on them. As his business was skyrocketing to success and the world was learning about luxury wine as well as the great Napa Valley through this legend, it appears that along the way he was constantly challenged with the decision to either grow the business much larger or pull the reigns back and keep it family-owned and boutique. As you most likely know, given the prominence and mass awareness of the Mondavi name in the wine industry today, he decided to grow and push it to millions of cases of wine. Then, in another milestone moment, he elected to have his company go public on the Nasdaq stock exchange, which was unprecedented at the time. I believe these decisions later placed great strains on the business and naturally defocused the management team in ways most wineries are not accustomed to. It was ultimately these choices that made him very financially successful personally, but most likely forced him to sell his business in the end, something I don't believe he ever wanted to do.

Planning for success and how to approach growth is imperative, and not only for making those critical future decisions. I strongly believe that this exercise requires you to "envision your success before it happens", which has proven to have a positive impact on any business as you start out and grow.

Let's reflect upon the very powerful words and deep encouragement offered by Bushwood Country Club golf professional Ty Webb to caddy Danny Noonan in the great movie *Caddyshack* as they discussed Danny's success while playing golf:

Ty: You've got to win this hole.
Danny: I thought winning wasn't important.
Ty: Me winning isn't! You do!
Danny: Great grammar.
Ty: See your future. Be...your future. Make...make it! Make your future, Danny.

Still gives me chills. I think.

As the business becomes more successful, you will have to answer this one simple question - "To scale or not to scale". Ride the wave to the peak and jump off even higher for an awesome photo-op moment, landing ready for the next massive wave. Or stay within the curl of the wave and safely ride it perfectly all the way through, still coming out with an excellent score. Is your plan to scale the business, either within the brand or with new brand offerings or acquisitions, or is it to stay the steady course, keep the business boutique, and strive for perfection in all ways? I wish I could tell you that making these decisions has repercussions that are as safe as surfing, like just diving into the wave and waiting for the next one. But given the massive complexities and risks associated with the wine business, the impact has serious reverberating effects that most times cannot be recovered, especially if the business was created with an alternative plan. Trying to shift a boutique wine strategy to a large-scale strategy is one of the industries riskiest moves and an effective way to have your wine business fail . . . and fail spectacularly.

However, I believe that intensely contemplating and knowing this today as you begin the journey creates a very different mindset, plan and approach for you and all involved. A large-scale wine business requires more capital, more time, essentially more everything, along with a different team skillset, rather than a wine business that has a goal to reign in and grow in a more controlled way. This doesn't mean it's not possible for one or the other scenario to be successful, but that it is unlikely based upon what history has taught us.

With this in mind, there are really two major strategies that seem to apply to the majority of wineries and wine brands around the world:

1. *Boutique, Niche Strategy*
   This is typically an isolated brand or winery, meaning there are usually no other items in the portfolio associated with the main brand and the volume is fairly small to mid-sized, which is always open for interpretation in the wine business, but generally means 50 to 100,000 cases. Also, the wine is not widely available, usually driven by the managements plan to keep it limited. Or it can also be that there are regional limitations, such as a winery focused in smaller areas like Temecula, CA or the Finger Lakes of upstate New York. Great famous examples of the boutique, niche strategy include Rochioli Vineyards & Winery in Sonoma, Colgin Cellars in Napa, Domain de la Romanée-Conti in

Burgundy, France, and Giacomo Conterno in Monforte d'Alba, Italy.

Also, this strategy could be a very focused, niche wine company like an online D2C négociant. There might be just one or many wines crafted under the same brand name but the brand personality is consistent, price points are relatively similar, and the overall brand imagery and experience is the same across the portfolio.

Most importantly, this strategy has a planned volume "ceiling", meaning that as success comes, there is a cap to how much they will sell or how many customers they can effectively service and entertain to sustain the overall strategy. Once at that point, this approach should be profitable, seamless and efficient. Then, most importantly, the team focuses time, resources, expertise, and capital on elevating the experience to higher levels of perfection, not building more volume of wine or increasing customer visitations.

Basically, success with this business model is measured in more than just volume of cases and dollars but in staying true to the original plan of being boutique and niche.

2. *Scalable Strategy*
This has some similarities to the boutique strategy, especially the luxury perception to the customer, because most wine brands want to be perceived as small and special. However, the goal is to scale the business to be much larger, either by increasing volumes of the flagship brand, launching secondary or more labels that may or may not have an association back to the flagship brand, broadening sales and distribution channels, and finding ways to increase visitation well beyond a boutique strategy. Basically, the sky is the limit. Great examples are Joseph Phelps Vineyards in Napa (Insignia & Joseph Phelps), Gaja Winery in Barbaresco, Italy (Gaja & Ca'Marcanda Margi by Gaja), just about every first and second growth Bordeaux producer (Châteaux Margaux & Pavillion Rouge de Châteaux Margaux). Other examples are

online empires such as Winc.com or the many négociant brands like Cameron Hughes Wine.

Here, success is measured upon how far and wide the business can reach, both in terms of wine volume and number of customers.

I understand there are many variations to the two paths because the wine business is never that simple. However, at the highest level, when creating the vision and plan for your wine business, these are the two directions that need to be decided upon and planned for before embarking down the road. And as you are making this very important decision, it's essential to know them well and deeply explore the pros and cons of each.

The Boutique, Niche Strategy

The goal here is to offer artisanal wines without many volume burdens to alter or affect the wine styles and quality, management decision-making or excellence of the customer experiences. This doesn't mean a scalable wine brand cannot achieve that prominence; however, capital requirements and decision-making within this wine culture and strategy come about differently. Since there are essentially less growth expectations, all the team's energy is placed squarely into acquiring the greatest wine and creating an exceptional wine experience down to the smallest detail. In addition, the sales and marketing strategy is on a more limited basis, which doesn't necessarily mean simpler, but an approach that is more controlled and narrower in scope. Let's distill this down to some concise thoughts about its pros and cons.

Pros:
- Simple, artisanal brand image that the customer can relate to and focus on. As the brand rises to the highest luxury level, the imagery is clear and associated with a single entity. This increases the brand's overall strength and makes it easier to create a focused marketing story and strategy.

- The brand has more limited customer exposure, which creates a sense of scarcity and desire. Below, I'll give you the "con" on this as well, but scarcity for a desired item is one of the greatest brand strengths you can experience in luxury marketing, regardless of the product.

- Sourcing, winemaking and cellar teams have a solid and clear direction. Pressures are focused on the details of crafting or finding the best wines rather than trying to increase volumes, which usually means lower quality as it grows.

- Sales strategy has less channel options to manage because the wine is usually created in limited supply to spread out across multiple avenues. Also, the direct-to-consumer (D2C) experience is more controlled, which promotes a simpler environment to create the super-unique imagery and experience in the winery or online.

- As I write this next point about "lower costs", I know what you'll be thinking, but we all know they are still generally very high because the luxury product business is never inexpensive. However, overall costs are less when operating a wine business of this size with very specific goals, but many factors can influence this, such as winery location, equipment, vineyard fruit costs, etc. With this strategy, there are fewer large-scale financial pressures, as opposed to the scalable luxury strategy which has to invest in increased cellar space and equipment to handle bigger lots, larger packaging requirements, off-site production, a more complex compliance approach, more skilled employees, etc.

- Finally, there is a focused energy from you and your team, thus creating a much more powerful brand image and customer experience, and usually the increased chances of offering a quality wine and providing an exceptional experience.

Cons:

- First and foremost is a financial matter – Put simply, the inability to increase the top-line revenue quickly and with less risk. To grow the revenue at a boutique, single-branded wine business, all the focus must be on either the sales channel mix - for example D2C vs. retail - and/or price increases. There are no other significant avenues to increase revenue once the brand "arrives" at need-to-have level.

- This is an interesting con because it pertains to the sourcing of the wine, internal winemaking, and cellar teams. I find they can get a little complacent and less motivated when crafting the same wine program over and over, year after year. There is always room for improvement regardless of how sought-after the brand becomes. But typically, wine improvements are long-term, subtle adjustments toward perfection. With the scalable, high-growth strategy, a secondary label, and/or a significantly higher volume goal of the main brand, the team's challenges are always new and you have more ongoing motivating tools to keep them on their toes.

- With luxury wines, vintages play a significant role. And depending on where the wine originated - Bordeaux vs. Napa for example - the secondary labels seem to have more latitude or 'forgiveness' from the critics, retail buyers and customers in difficult years. With a boutique luxury brand strategy, the vintage quality is either on point or not on point. There is no option to move this fruit to a more affordable secondary brand. Either you make the wine with the compromised fruit and hope for the best, or you reduce the number of cases produced that year and sell it off quietly on the bulk market. I've seen both scenarios, and sometimes decisions are made based on financial pressures, which isn't optimal in terms of either wine quality or supply.

- Let's say there are mostly great quality vintages, which is common in Napa and Sonoma. The issue then becomes the ability for the winemaking team to alter the blends to perfection. When there are other brands to "lean on", the team can move juice around for the sole purpose of dialing in the final blends and style. It's like a chef being able to have a kitchen full of great ingredients but not needing to use them all because the farms will pick up the lower quality food, which of course never happens. Having blending options allows for pure artistry to take over in the final wine. If this chef had to use all the food in the pantry, not everything would be of the highest excellence, unless he or she were willing to throw some away.

- With a limited amount of cases to sell, once your business plan reaches its goal, not all preferred sales channels can be satisfied. Now I understand it is desirable to have a "hard to get" brand, but with the wine-seeking customer, if they cannot obtain your wine at some point, they will quickly move on to another brand that can scratch their itch, leaving yours behind. This also may drive up the prices. It's good to be untouchable in perception, but touchable to those customers that you need.

Though there are many excellent aspects of the boutique model, it is important to weigh the associated risks of decreased options of revenue scalability, challenging team motivation methods, more limited winemaking options, and potentially not being able to satisfy every targeted luxury wine customer.

The Scalable Strategy

This long-term strategy is to have a wine business that has size in volume along with a broader distribution. The goal here is to make the great wine and provide the best experience possible, but the team must also balance that desire with other higher growth and sister-brand, or second label, requirements and tasks. The sourcing and winemaking team has more pressure but also more flexibility and options at their disposal. Also, the sales and marketing strategies become more complex and broader, managing multiple labels and more expanded distribution requirements. Let's distill this down to some concise thoughts on the pros and cons.

34

Pros:

- The first and most obvious appeal is top-line revenue and overall growth potential. When done well, this additional capital investment allows the business side of the winery to have more options to drive growth through buying other wineries or vineyards, hiring more experienced people, and buying ads on social media to expand clicks and broaden the business into other categories.

- It mitigates the risk of making it through bad vintages. In Bordeaux, if there is a massive hailstorm in early spring which takes out a big chunk of their fruit and compromises the quality of the rest, the critics and the fine wine press make this news story very clear to the buyers and consumers. Scores go down and feature articles are written, thus reducing the demand of that particular vintage. Usually the Châteaux's second labels get more focus, not only in the winery but also with the consumer who wants to stay within the brand but doesn't want to pay the exorbitant prices for lesser quality wine at the Grand Vin or flagship level. So, with a larger portfolio, vintages become less important than the style and consistency.

- Having a larger revenue stream and subsequent cash flow (you hope) helps with investments that grow the brand name. This opens up many options for the future, including better marketing partnership associations, vineyard selection, wine brand acquisitions, or the ability to hire more experienced talent. It can even help you launch into other categories like hotels or fine cuisine, or simply provide a cushion for any potential down years. Basically, more cash flow creates more options . . . as long as the main brand's perception is not damaged along the way. It's brilliant when it all comes together.

- The larger, more challenging portfolio gives the sourcing, winemaking and cellar teams more interest, excitement and more dynamic goals to keep their engagement levels higher. It also gives the sales and marketing team more portfolio options to sell in expanded sales environments.

- The following is a very important point, and something I've seen happen a few times in my career. It offers your customers a more approachable path to your top wines, assuming you have wines with higher price points. Follow me on this one. It's clear that with any luxury product, and at times with many less expensive products, as the customer becomes more educated about the brand and more familiar with the company that creates it, a percentage of that customer base will buy "up the chain." An excellent example is golf balls. Titleist has product lines ranging from entry-level to prestige-level golf balls. They target their entry-level balls to the new golfer and as that customer becomes more sophisticated at golf, they sell the idea that the next level golf ball will improve their playing experience. The customer feels that Titleist got them this far, so it is best to stay with the brand and buy more up the chain.

The same holds true for wine. As the wine customer is learning, though not yet willing to pay extra for a more luxury wine brand, the second label or a higher-priced wine in your portfolio offers them an option to explore your brand at a different, more superior level. As the relationship with your customer builds over time, so does the opportunity to sell them "up the chain". And a percentage of customers will go there if you market to them correctly.

An excellent example that comes to mind is the Quintessa Winery located on the Silverado Trail in Napa, and their second label, Faust. I recall when Faust was first launched. It was made clear that the same team crafted it with fruit mostly from the same property. Quintessa's main brand was $110 a bottle. Faust was $50-$60 a bottle and was of excellent quality. However, Faust quickly became a D2C and retail darling. Sales representatives for the company were invited into retail accounts they would have never been part of before. And as Faust grew, so did the awareness and sales of their top wine, Quintessa. If a representative is selling Faust by-the-glass to a

2-star restaurant, they can more easily ask the buyer to add Quintessa to the by-the-bottle menu. Many customers found the flagship wine through a by-the-glass Faust at Wally's in Beverly Hills, for example. It was and still is brilliant execution, and the whole fleet ascended with the rise of that tide.

A scalable wine brand strategy has many very appealing positive aspects, such as financial growth to support other business opportunities, more wine-blending options, and a path for non-current customers to find the flagship brand.

Cons:

- The most important con, and by far the riskiest, is the very real possibility that the luxury perception of your wine business is damaged due to larger volumes and possibly lesser quality. Not to come across as a drama-queen, but this can be catastrophic. Usually the damage does not become apparent until you've invested heavily in your growth plan, the wine, the business, and the team, and the second label or portfolio additions are well into distribution. It is a slippery slope because your strategy calls for growth, so it is necessary to go down this path and grow. But if you do not achieve the goals, you are already committed, and back-tracking takes years and a significant investment. At times, brands never recover. It's no secret that Robert Mondavi's flagship brand suffered, and I believe still does to this day, because of the launch and association with the lesser-priced, more volume-minded Woodbridge.

  Another example is one I know intimately. As I was managing the prestige sparkling wine house Mumm Napa in Rutherford, Napa Valley, I was told that G.H. Mumm Champagne, our flagship winery and brand, was at one time the #1 imported wine from France to the US. Based on the placements alone in the early black-and-white movies, I believe this was true. In a scene from Casablanca, Humphrey Bogart and Ingrid Bergman sit romantically at a small café table sipping Champagne. There is a bottle of the legendary G.H. Mumm

Champagne Cordon Rouge in clear sight right next to them. It was iconic. Fast forward many years later. The company that owned G.H. Mumm Champagne wanted to broaden the brand and use the strength of the Mumm name to grow into the US market. They did this by building a baby brand called Domain Mumm. It was available at a much lower price point and exhibited a completely different but outstanding quality and style profile. People were visiting the beautiful new Domain Mumm (now Mumm Napa) in the US, something they couldn't do easily by flying to Champagne, France. Domain Mumm was much more accessible to the masses. Therefore, from this perceivably smart US investment, the flagship brand G.H. Mumm Champagne started to dissipate out of the US market, unfairly experiencing price pressures, replacements, and, ultimately, a damaged brand image. It's taken the company many years to recover, re-launch, and re-establish G.H. Mumm Champagne back into the marketplace. It is one of my favorite champagnes in the world and I'm confident it will find itself back on top, but at a significant cost and learning experience mostly attributed to aggressive growth goals.

- There is a very real financial risk with any increase in volume. Costs go up across the board, especially in production, winemaking and packaging. Experienced salespeople at this level of selling volume are typically the most expensive individuals in the industry. Additionally, making sure they have the marketing budgets and price discounting tools to be successful requires a serious investment of capital. Basically, it's a very intense financial outlay to take on the risk of building a growth wine business.

- As the volume and portfolio grows, it's imperative to have a solid sales and marketing plan in place. Without it, two things typically happen in sales:

    1. The team will "default" to the lower-priced items in your portfolio if they experience even the slightest buyer push-back on the more expensive wines. This

could become a slippery slope because salespeople are hunter by their very nature, and as such, they need to bring something back from the day's hunt, which could be at more deeply discounted prices than the company strategy calls for.

2. Secondly, if the goal is to scale the growth of your most prestigious wine upward, then the risk becomes that the wine is sold into undesirable channels that denigrates the brand image. To achieve higher volumes, depending upon how large the goals are, there is a natural pecking order in retail of where to go and what to do from a marketing support dollars standpoint. To sell more wine, it needs to be pitched to retailers that can move that level of volume. Or, similarly, if the wine business is online D2C, the methods would be to offer significant discounts. And to be in that larger lake where more sophisticated wine companies swim, especially coming from your little pond, it requires a higher level of sales expertise, management and finesse. The result is you will need to increase marketing support dollars to get it placed properly, or if in a D2C model offer free or discounted shipping. As closely as you try to manage this, it can easily get out of your control in areas such as discounting, placements, associations, and quality, like shipping D2C wine in hot delivery trucks in Florida during the summer. Be careful not to let your brand image slide away from the core strategy and perception. It is the golden goose.

- If you build your wine business based upon the boutique strategy, then shift to the scalable strategy, your existing team most likely will not have the skillset to make the transition. Selling boutique wine to a much smaller D2C and retail-base requires very different levels of team experience than selling a brand that is scaling to a broader online presence and larger retailers. There are very few sales and marketing representatives that have the skills to sell and manage a 5k case luxury brand and a 50k case brand at the same time. I won't bore you with the list of reasons why, but approach is different, contacts are different, wine knowledge is different, etc. So, a significant risk is your team skillset. Difficult decisions must be made to make sure you have the right people doing the right things in the right places as you scale. Not easy.

- As the brands are transitioning from small to larger volumes, the online presence as well as the hospitality experiences have to be well thought-out and planned. I strongly believe the customer that is only interested in the $50 level of wine will require a different experience and sales approach than the $110 level of luxury wine. If your customer is already at the $110 level, your online experience for them, the emails you send to them, and the in-person experience in the winery you provide to them is a much different animal than what you need for the $50 customer. At the highest level, your goal is to strengthen the already solid bond between the brand and the customer, and this requires other like-brand associations and experiences. However, if you are in growth mode, this requires a different way to associate with these customers, which can spread your team out or dilute the marketing effort to each customer.

Though the scalable strategy has many appealing aspects, it's important to be very knowledgeable and intimately aware of the many risks. You could potentially damage the flagship brand image and also increase your financial obligations, which always pushes the business into areas not originally intended.

As a final note on this very important question of "to scale or not to scale", the business model strategy you select will depend upon many factors, some of the most important being availability of financing, access to capital, your aptitude to handle the unknown moments that require you to spend more of that capital than you may have anticipated, access to operational facilities, finding the right talent for the goals, and hiring a significantly experienced team. And though these are extremely valid issues to resolve, I believe the final answer will come to you quite easily by answering this final simple question

- From your heart, what do you ultimately want out of the wine business? Is to it to conquer the wine world or live quietly on a vineyard-covered hill in Sonoma, CA? Is it to have a multi-national wine empire or the next cult wine? Is it to have the finest online D2C wine club or a personally satisfying small winery that has a high level of hospitality? Answer this and you will have made one of the single most essential steps toward your dreams. Regardless of the plan you select, knowing it now and sticking to it when important decisions need to be made while in those times of great accomplishment will be the difference between your wine businesses ultimate success . . . or its disappointing and costly failure.

10 Imperatives To Triumph In Wine

Purposely, I've spent a fair amount of time having you do some deep soul-searching about your future direction and strategy. Both of those methods have proven to be highly successful to those that have understood their DNA and stuck to the original plan. Of course, there are those many examples that didn't work out as planned, devastating family legacies and personal fortunes or seriously disappointing investors. But you are now ahead of the game and have significantly minimized the risk of becoming one of those unfortunate stories. Kudos to you!

Since we're on the topic of *things to know and embody before you start building your wine business*, there are a few other highly important imperatives you will want to keep in mind as your plan begins to unfold.

Here are ten imperative steps to take toward your goal, regardless of which direction you choose. They apply to both new ventures as well as reboots of existing ones. Just kindly bend your mind a bit to apply these steps to either.

1. ***Start with a plan***
   As we've covered, possibly in more detail than you wanted, many entrepreneurs start with just the wine and not much else, believing

this is all they need. And if you start this way, you're starting the process to fail. Those few that succeed start with a solid, well-thought-out plan. In addition, it would benefit you and others greatly if you create 1, 5 and 10-year sales and marketing plans with clear goals, realistic timeframes toward milestones, and an associated financial plan to justify them. I don't want you to spend a lot of time over-planning, but the point of having something solid in place before crafting the wine will set you in the right direction.

2. *Find the best and most experienced talent available*
The wine industry is sexy, there's no way around it. I'm assuming it attracted you the first time you visited, as it has done with thousands of extremely talented, highly-intelligent and über-capable people that call it a career. The industry not only has endless passion, but it has talent and a lot of it. Armed with your Passion Statement, plan and purpose, surround yourself with as many of these great people as you can. Once on your team, they will go to war with you and be loyal for as long as you wish to inspire and lead them correctly.

3. *Have buy-in at all internal levels*
Most decisions happen in the "ivory tower" then come down on the team. Selling the vision and how everyone's careers and livelihoods will improve is as important as the execution of the plan. If the team doesn't buy in, and I mean buy ALL in with the purpose, passion and how they will make the impact happen, your plan is a ship without a rudder for sure. The team is your conduit of the strategy, so having them agree with it, if not more so than you, is imperative.

4. *Align the team's talents with the plan and goals*
Asking a boutique winemaker to go from 5k to 50k cases with the same level of quality is tough. Asking a wine sales representative that originally focused on Michelin-starred restaurants to then sell to larger retailers is asking too much. Similarly, asking a boutique wine brand director to build marketing plans for a regional restaurant chain is usually a no-win proposition. The message doesn't always translate, and often the talents of your team don't

always translate. Each position you either hire or is already in place needs to be realistically evaluated for their abilities. You can determine who makes the transition and who needs to be repositioned. This doesn't mean they are not talented at their craft, they're just not a perfect fit for what you are looking to achieve moving forward. There should always be room for outstanding and loyal talent. Finding a happy place for your best people to be included and empowered is important.

5. *Have a clear customer message that the team knows perfectly*
Regardless of the strategy you select, the overall message needs to be simple, clear and concise. It should be a business statement that you can express to someone in 30 seconds and finish with them fully understanding the finer points . . . because sometimes your sales team is only given 30 seconds of a wine buyer's focus in a new item presentation meeting or pouring samples at a tasting event. And more importantly, the team needs to not only know the message but practice it with you to guarantee that it is perfect.

This last simple point is very important. I've built a few very strong messaging platforms for a variety of wine brands and I've had incredibly experienced sales and marketing professionals look me in the eye and say, "Eric, great stuff. I got this solid. The message will be delivered perfectly." Then I attend a few customer meetings with them and watch how the brand messaging gets tossed about, changed, jumbled, or not even delivered in a few instances. After learning and seeing this firsthand, I've always made it an exercise for the team to practice often in front of me as well as in sales meetings with fellow associates; pressure moments similar to a buyer meeting. It was then that I saw the message being delivered perfectly and consistently to the customer in each sales scenario. It's a very worthy practice.

6. *Protect the brand image at all costs*
At whatever cost, protecting the brand image, essence and associations is paramount. It is the sole reason for everything to exist. You may have outstanding wine, a 30-million-dollar winery, and incredible talent, but if the brand image slips, so does everything else. This statement is mostly directed at those

decisions many management teams make that chase dollars or percentage growth goals at the expense of the brand image. Every decision you make starts with the question "What will this do to our brand image?" If the answer is anything other than building it, it's most likely not the right decision.

7. ***Commit to world-class wine . . . and support it***
Regardless if the goal is to keep it boutique or scale larger and expand, there needs to be a high confidence level by you and the winemaking team that is firmly supported by the financial plan. With better equipment, additional reliable sourcing or winery production (people, space and equipment), the wine can be crafted at the highest level possible, regardless of whether it is $8 or $800 per bottle. Nothing is compromised. If there is any doubt amongst the team or customers, it needs to be hashed out before the engine starts. The sous chefs in the kitchen of a 3-star restaurant know that the ingredients provided to them by management are superior, and it's their job to take them and translate them to the customer experience at the highest-level. If they are directed to create 3-star cuisine with lesser ingredients, the business and culture as a whole will suffer, and ultimately this will translate to the customer and the bottom line. So, promise to be excellent in all ways, not just verbally.

8. ***Commit to world-class service . . . and support it***
As we've learned by now, great wine is not enough. It is needed and helps immensely when the wine is really great, but wine alone will not get you to your goals. As you will read further ahead in a section dedicated solely to the noble ideals of instilling luxury and excellence into the organization, committing to having world-class service is a game-changer. Firstly, there are so many wine brands and wineries available to your customer that offer excellent service, it basically puts your wine company on par with those competitors to just stay in the game. However, instilling world-class sales, marketing and hospitality services at all levels will be the deciding factor between failure and great success. And it will need constant repeating, development, nurturing and investment to stay at the highest level required. Think of it as something precious in your life that demands constant attention. Without it,

it goes away quickly. With it, everything is lifted to a higher level of satisfaction and success.

9. *In wine, slow and steady wins most races*
I've been involved in situations where steps 1 through 8 above were handled with excellence for long periods of time. Then, many times, some outside pressures emerge and everything that was built starts to unravel. This is usually due to the over-exuberance of the owner or management in charge. Sometimes they are so confident in the plan (or more fittingly, themselves), or possibly there are financial pressures outside of their control, they don't consider the fact that crafting and selling great wine at a superior level takes time for an endless array of reasons. If the expectation of a realistic timeframe is factored in, the team buys in more completely and the process is considerably smoother. There is more upside to overperform and the quality of the message and wine is exponentially better. Also, you have more time to shift if the strategy doesn't unfold as planned, which it rarely does.

Believe me, I understand that the indulgence of time is not always a possibility. If this is the case, I would strongly advise to start back at square one and factor in the realities because without time, the plan will never unfold in the way you envision. Having the time to not just perform tasks but accomplish them at the highest level of excellence needs to be an imperative. Otherwise, you're building a luxury high-rise using toothpicks.

10. *Never rest on your success*
The wine business moves fast and becomes very busy, especially if you find success along the way. A million and one distractions, both necessary and unnecessary, buzz around, constantly sucking up valuable time and energy. What usually gets left behind is innovation and exploration. Even for boutique strategies that have an outstanding wine and world-class service, it is important to constantly push excellence forward. Experiment with new methods to make the wine, even if it is subtle and takes considerable time. Try new ways to engage with your customer and buyers down to the smallest of details. If you read the wine

industry publications, you will invariably come across little nuggets of examples every so often. You might read how a well-established, super luxury winery team in legendary Bordeaux Chateaus test new vineyard techniques on small blocks of their estate. Or take a look at any retail shelf and notice how some of the brands will make subtle changes to their label designs. There are numerous other examples, all worthy efforts. Rest when you retire; until then, continue to innovate, inspire and experiment.

This is a simple, 10-step guide intended to give you a brief yet comprehensive way to plan the most important aspects of your wine business. So plan and plan well. But at some point, put a stake in the ground and turn that plan into action because the wine industry doesn't wait for those who wait.

SECTION IV: LUXURY & THE ART OF EXCELLENCE

*"Wine offers a greater range for
enjoyment and appreciation than
possibly any other purely sensory
thing which may be purchased."*
– Ernest Hemingway.

Before we go into the finer details of wine sales and marketing in the following sections, I think it's imperative we spend a few moments getting an understanding of what the terms *luxury* and *excellence* mean to the wine business, and your wine venture in particular. It is important to know that at either $9 or $900 a bottle, wine is not a necessity of life, even though it might seem that way after a rough day. For this reason, consumers perceive wine as a luxury product regardless of the price point.

The legendary French fashion designer and businesswoman Coco Chanel really understood it best when she said:

Luxury is a necessity that begins where necessity ends.

You are in a luxury business regardless of wine region, price point, style of wine, or how spectacular the architecture of your winery may be. And given the significant amount of complexity and details in getting a wine into a customer's glass, personifying excellence is an absolute requirement to create wine at any level of success. This needs to be embraced by those of us in the industry who strive to sell and market wine in the most effective ways possible.

To appreciate this further, we need to answer the question, *"What is luxury* and *what is excellence?"* You can have excellence without luxury, as in a surfer riding the world's largest wave and handling it perfectly. But you cannot have luxury without excellence because it is just a marginal effort and a lot of money spent finding out the hard way. Regardless of the path taken, one outright fact is the practice of striving for excellence on all levels is an absolute must-have. Without it, this guide or any other business guides will be meaningless. So let's start by first defining and getting on the same page of what *luxury wine* means. Then we will embark upon the noble ideal of the *pursuit of excellence*.

The Many Perceptions Of Luxury

I use the term *luxury wine* as opposed to prestige wine, fine wine or any of the other well-used terms because, to me, luxury defines a complete experience, not just a product. As you read further, you will find that the complete experience is what will separate you from the countless other wine brands and vineyard-covered hills. If you search the Internet, the definition of luxury seems to be fairly common across the board. Where I found the difference is in the application of the term. I mostly favor the Cambridge Dictionary version, because who can disagree with the people at Cambridge, right?

> ***Cambridge Dictionary Definition of Luxury:***
> *Great comfort, esp. as provided by expensive and beautiful possessions, surroundings, or food, or something enjoyable and often expensive but not necessary.*

I think we can all agree on this definition. Pretty straightforward. However, can we all agree on its interpretation to each wine consumer? Consider the following:

Luxury means different things to different people. It could be a weekly gathering in a simple local restaurant for a modest family living in a less-fortunate country, or, conversely, it could be an evening in a beautiful mansion located on the cliffs of Positano, Italy, for an über-wealthy family. For my mother who grew up with Gallo burgundy wines, luxury is a nice $10 chardonnay with an ice cube keeping it chilled for an hour while she sits in her garden out back. For me, luxury is sipping, from a Riedel Vitis flute, a G.H. Mumm Champagne Mumm De Cramant Grand Cru while sitting on the cherry red couch of the La Maison Mumm in Reims, France. For us all, the opinion of our personal luxury experience is influenced by the product, the people, the surroundings and the emotion felt. There is a perception of a high level of excellence as well as an emotional attachment that we feel, however we define it.

I think it is important to focus on the term *perception* as we explore an understanding of the customer and the luxury experience. Referring back to one of the earlier comments, Walt Disney tirelessly focused on providing an outstanding experience and product. He wanted it curated to the point that his team knew that when they were onstage and out in front of the customer, it was imperative to create and impart the Disney magic. Designing a perception and full immersion of pure fantasy was as important as the quality of the rides and theme parks themselves.

Let's narrow this to the world of wine. At this very moment, I'm going to ask you to change the way you think about the product you are crafting and selling, a paradigm shift that won't be easy as it has been ingrained in wine country for centuries. This is not just about excellent wine and a nice label. With the thousands of wine brands providing similar visual experiences, bottles, wines, and so on, it has to be *so much more.* This is about taking an inessential product and creating the perception that it is an essential experience in your customers' lives. Your aim is to create the highest brand value by leveraging all intangible elements of uniqueness, such as place, people, tradition, origin, artistry, skill, scarcity, prestigious clientele, etc.

You're not creating and selling wine with a cool label . . . you're selling an idea, an emotion, a lifestyle.

Defining Luxury Wine

Here is a statement specific to defining luxury wine, taking into account all the various and important aspects of the wine experience. Please use it as it is, or feel free to tailor it to your specific situation and business model.

> *Luxury wine is the curating of an exclusive, evolving, unique experience that revels in excellence with the attention to detail, offers meticulous and superior craftsmanship of wine and service, and delivers unexpected indulgences, all in a perfectly designed atmosphere or packaging.*

Please reread through that statement a few times and pay close attention to each word. They all have a place and specific meaning.

- ***Curating***
  Crafting a complete luxury experience in a personalized way.

- ***Exclusive***
  As you will see below, the luxury customer thrives off of feeling exclusive and special, so finding a way to emanate this is important.

- ***Evolving***
  To get to the top is one thing, to stay there is a whole other task which requires you to constantly evolve the experience, staying true to the brand and winery's personality. A great way to see this is to search the Internet for the evolution of the Bentley. It changed with time but stayed true to its origins and purpose.

- ***Unique***
  The experience cannot be the same as the other 10 luxury wineries down the road or 100 wine bottles sitting next to yours. If it is, yours is easily forgettable and replaceable regardless of how great the experience, wine or label is. Be unique. You must offer or exemplify something special that

the customer will only associate with you, your wine and your winery.

- ***Meticulous & Superior Workmanship***
  It goes without saying that the wine and packaging needs to be of the highest pedigree. This doesn't necessarily refer to label bling, scores or ratings, but the idea of luxury wine dictates that you need a far superior product than most available at similar price points. This also applies for the personalized service on the phone, online and in the winery. Superior workmanship should be a part of everything you and your team do every day.

- ***Unexpected Indulgences***
  Luxury wine is an inessential luxury. No one loses sleep because they didn't drink excellent wine that day, though from what I'm told, I can get very grumpy if I don't. It is an indulgence of the senses and emotions for those who seek its charm. But focus on the word *unexpected*. Your experience needs to give the customer the Aha! moment, and an unanticipated response.

- ***Perfectly Designed***
  The definition is finished off with a term that evokes excellence in the wine, packaging and experience. Be perfect in design and excellent in execution.

Luxury wine is the total experience from beginning to end; all people, all places, all services, all products, all everything… *all-in*.

Making Excellence Culturally Imperative

The world of wine boasts countless outstanding designs, wines and wineries. One after another are stunning. All too often, and especially at the luxury levels, we focus on excellence in wine, the artwork on the label, and the architecturally pleasing way the firm designed the view from the tasting room. But what about the special relationships with the vendors that assist in putting it all together, or the retailers that buy the wine for their restaurant clientele, or the visitor partnerships that bring people to the winery in the first place, or the training your sales representatives require in handling special circumstances? Going even deeper, what about the quality of the backend systems needed to manage such a complex business, or the investment in employee wellness and career development? As an industry, we can't say that the same level of investment and commitment to excellence is applied to all parts of the business as it is with the wine. And I'm not just referring to the cost of a sales and marketing employee, which is usually high, but also the investment in educating, training and empowering that person to achieve excellence. Sure, give them the wine and the place, but in today's wine world, that is just not enough.

There is a culture of excellence in the winemaking and wineries themselves, but is there a culture of excellence in all parts of sales and marketing? I don't believe so.

To be a serious player in any luxury craft requires a commitment and philosophy of excellence that starts at the top and is embodied all the way to the lowest-paid employee. A sense of unconditional passion and a desire for pure excellence must exist onstage in front of the customer, and offstage behind the scenes, in all parts of the winery and company. From the private room where the tasting is held to the bathrooms the customers use on the way out. From the email sent about wine allocation to the social media posts, a commitment to excellence must reign supreme in all parts, in all places, all the time. Your team and your experience need to be the embodiment of excellence at all levels, particularly in sales and marketing.

However, the very notion of attaining perfection is not actually achievable. There is no end line. It is a never-ending moving target. Each time you get closer, the line moves further away. It's an exasperating proposition, but that is the beauty of it as well. As driven, passionate, highly educated people we need this challenge to feel accomplished and alive. And, to set any luxury wine business apart from the rest, it is an absolute requirement.

"Perfection is not attainable, but if we chase perfection
relentlessly, we can catch excellence."
— Vince Lombardi

As you read through the sales and marketing topics in the upcoming sections, imagine how you can apply that sense of excellence with every task performed. From a simple customer email, to a small handout used in a partnership, to your daily social media post. Make that commitment to yourself and empower your team with the same awareness of importance and determination.

Case Study: Master Classes On Film

Time for a little homework.

There are a few documentaries that have had a profound effect on how you can approach the management of wine. These restauranteurs take the term excellence to another level of meaning. As a next step, close this book and watch the following movies.

Jiro Dreams Of Sushi
A master class on the never-ending pursuit of
excellence using traditional methods.

Chef's Table: Season 2, Episode 1, Grant Achatz
In a more modern setting, another master class on
the struggles and artistry of excellence in
innovation and reinvention.

After watching, you may feel, as I did, that you may have done very little compared to the relentless passion and pure awesome creativity these people possess, and that's okay because it's all about learning, starting new habits and imparting new beliefs. You will find these stories empowering and motivating beyond imagination. Watch closely how they pursue brilliance and notice how it can be absolutely replicated in your wine venture. Use these messages of unending passion and pursuit of excellence to inspire your own journey. The road leading toward perfection in every detail is an imperative that starts from the top and can be part of the cultural fabric of your wine venture.

SECTION V: UNDERSTANDING YOUR CUSTOMER

"The customer's perception… is your reality."
- Unknown

Now that you have some of the major pieces in place, the next step is of course to focus 100% of your time and energy on the wine, label design and winery, right?

Not yet.

The next step is to become very knowledgeable about your future customer. These are the people that will be deciding if you survive or fail, and they will also dictate the style of wine, winery, packaging and overall experiences you curate and offer. For these reasons, it makes pretty good sense that this should be at the top of your list of things to think about and know well. To align either the creation or the repositioning of your wine business with that of the wine consumer, we must first understand who these nice people are and what interests them. Only then will we understand why they buy wine.

Upfront you should know that my deductions are not based purely on statistical data and research. You can find droves of wine data in many places online, or in books written by university staff. And, quite candidly, I find it somewhat disappointing that many smart businesspeople and wine scholars put such a large emphasis on data-only plans. Data collecting becomes applicable and useful only when street smarts, experience and intuition are generously sprinkled in.

I've compiled these general profiles from combing through years of that tedious research myself, then filtering it through personal and business experiences from directly managing many wineries and wine brands. I've had the unique opportunity to closely watch these brands as they were distributed out to the marketplace at many different price points and in all the various sales channels, such as high-end wine shops, D2C online, large retailers, and white tablecloth luxury restaurants, to name just a few.

Also, at the risk of displeasing some wine industry scholars and experienced wine marketing professionals, I should mention that this is a general profiling of the average wine consumer regardless of the price point or style of the wine. Sure, luxury wine geeks may buy much higher-priced wines in different places than that of my mom who buys wine in her local grocery store, but I believe the reasons for buying wine are very similar for the majority of wine consumers. As with any demographic customer profiling, there are plus and minus variations and generalizations, but the high percentages fit this bill.

Enough of the disclaimer. Let's move on.

Who Are They?

So, who are these nice people that buy our wines? They are typically:

- ***Between 30-55 years old.***
  As people mature and become more educated about the finer things in life, as well as have more life experiences, it is usually in their 30's where they will come across some better-quality wines and have an increase interest in learning about the details. Some discover wine later on in life, but I find the majority begin to "let it into their personal ecosystem" around this timeframe.

  Also, you may have noticed that I did not add in a 21 to 30-year-old category. If you look purely at the data and don't pay attention to the overt media noise about the wine consumption habits of millennials, roughly 70% of wine buyers fit into the 30 to 55 bucket.

- ***Higher than average discretionary income***
  To indulge in the luxury goods lifestyle, which wine is identified included in, there needs to be a higher annual income than the national averages. I've seen research try to quantify this, but I don't believe in placing a dollar value on these customers because it largely depends on where in the country they live. We all know buying a house and living in

San Francisco is a considerably more costly endeavor than living in Albuquerque. Discretionary income is the primary data point to focus on. Wine consumers have the income and desire to purchase products that are not essential to human existence, regardless of whether they make $50k or $500k per year. They may purchase different wines, but they all buy wine. When I was with large wine companies, we marketed and sold differently to different parts of the country. But at the core, regardless of where they lived, the majority of these consumers all purchased wine because they had discretionary income.

- *It is a couples thing*
  Especially as it pertains to visiting wine country where most of the wine experiences happen, couples love to discover wine together. It is a very romantic product, notion and activity - thus the couples thing. They drink wine in wine country, and they drink wine at their favorite restaurant. Again, that could be a local Italian hangout or an upscale, urban, fine-dining French bistro. You get it.

- *50/50 female and male, but the women drive the decision*
  Lower end price points skew more towards women since they frequent the places that wine is more often found, like grocery stores. As the per bottle price point increases, more men enter into the picture and usually top out at 50% of the total. However, I believe the couples-thing plays heavily into this. Most wine purchasing decisions are made with the significant other in mind. For example, I've experienced that when a couple visit a winery, he might be the reason they are there. However, most of the time, he checks with her on the wines to purchase, making sure it's a wine they both will enjoy. It usually happens as a whisper across the tasting table right before purchasing, but I believe she drives the bus with most wine decisions. Something data doesn't reflect if you are looking at just the numbers.

- ***Well-educated and seeking to learn more***
  Again, a higher level of discretionary income usually comes along with a higher level of education like a master's degrees, or the professional training it takes to become a lawyer, doctor, executive, and so on. But they don't stop here. They are thirsty for more knowledge because their years in education institutions conditioned them to constantly seek understanding and knowledge. And wine is perfect for this mindset. Regardless of whether it is a simple French rosé or a monster Napa Cab, the process of getting from the vineyard to the glass is complex, interesting and endlessly fascinating to the wine customer. It is an ocean of learning.

Again, this is not a cookie-cutter formula. There are variations for days. However, this profile captures the majority of your wine consumers, providing you with an excellent understanding of who they are.

What They Do & Why They Buy Wine

Now that we have that understanding, let's explore what they spend their time doing.

- ***They are social***
  As opposed to other types of alcohol, your future customer typically drinks wine with others in social settings like restaurants, events or dinner parties. And, though I don't believe the majority fully understand proper wine pairings, they usually have better foods with their wine.

- ***They love finer foods***
  As a follow-up to the previous point about being social, they have a greater understanding of finer foods. However, the word *finer* means different things to different people. It could be as prestigious as being a highly-capable home chef cooking restaurant-level dishes, or a less sophisticated person who enjoys a nice selection of everyday foods like steaks or pastas

made with more complex ingredients. Either way, overall, they enjoy the idea of knowing and having better food in their lives, and wine plays into this lifestyle perfectly.

- ***They are highly-skilled and usually accomplished at a particular profession***
  Most are confident and driven. This lends itself toward a more accomplished category of people. They spend a great amount of time being focused on and working on their craft, which could be a doctor, an outstanding teacher, or a highly-accomplished company executive. Regardless of the profession, they care about their careers and want to do well. Laziness is not usually a characteristic of this demographic.

- ***They enjoy travel and exploration, at a more enjoyable pace***
  They love to travel, as long as it's in comfort and style. They like the finer things in life and have the discretionary income to afford it. If they go on a trip, it's usually at a higher level of cost and refinement than how most travel. And while they are there, they will seek out the best places to experience life to its fullest.

- ***They return to places they love***
  There is a high level of loyalty to places and people they feel connected to. If they've been there before and loved it, they will frequent the same places when they return because they know what they like and it's important for them to indulge in it more often than once.

- ***They relish gaining special knowledge***
  Interestingly, they love to gain knowledge of how things work, especially items they love. And they like to believe it's an "inner circle" knowledge they can share with others later on, making them the person in the know. Think back to a time when a friend was showing you their new car. Very rarely is it, "Well, here it is, Bob, blue and fast." They expound about the year, the interior, how it was painted, how many were produced, where they bought it . . . almost every time. And the same holds true for wine. They want to express their inner

knowledge of the brand, the winery and the wine. It feeds the
ego and makes them feel special. They love to share, which is
important to note when you're building your customer base.
Referrals can be significant.

- ***They find wines through recommendations***
  Here is another interesting attribute that transcends all price
  points and styles of wine. They find wines though others they
  believe have wine knowledge and expertise. This could be
  from an online wine blog, a wine shop representative, a
  sommelier in a restaurant, or a friend who is also into wine.
  Remember, your future wine customer is constantly seeking
  new knowledge about the things that interest them, and wine
  is no exception. However, once they have the
  recommendation, they will make the final decision if they like
  it and if they will buy it, or not.

- ***They have strong family values***
  As they have matured and become more exposed and aware of
  the world they live in, they feel fortunate to have opportunities
  others don't necessarily possess. With this comes
  responsivities to others, either within their families or those
  that are less privileged. They care for their own and give with
  open hearts and beautiful intentions. Sharing and caring is a
  core value they strongly believe in and it is a necessity to their
  existence and happiness.

- ***Most importantly, they need to feel special***
  This simple statement is profound within itself. They are
  frequently treated well because most of the places they visit
  have a heightened level of superior service. But the intangibles
  that make them feel special are what they love the most. For
  example, there are twenty restaurants in NYC that a wealthy
  hedge fund executive can visit, all with excellent food and
  experiences. But the restaurant with the owner who knows the
  name of this customer and who greets him personally every
  time he comes in is the one that gets the majority of his
  business, hands-down.

In a few short paragraphs, you have become much more familiar with your future customer. Now is where the rubber hits the road, or where the cork is popped from the bottle. (Not as good, I know. I'm still working on it.)

The final question is, "Why do they buy wine?"

They buy and enjoy wine because it is a special indulgence that makes them feel they are living life at a higher level; that this simple glass of grape juice takes them from whatever stresses they may have that day to a place of acceptance and gratification; that it transports their minds to places they dream of being, or simply makes the moment with family and friends more relaxed and pleasing. Wine is special to them and thus they feel special when enjoying it. It's a powerful and very personal experience. Wine is their reward for living life.

The luxury of wine, at all levels, embodies the belief systems and habits of this customer demographic perfectly. Wine is one of the highest levels of perceived luxury, with excellence in service and craft. It's an acquired knowledge over time that few possess, and it can be shared with others. It can be experienced in public and at home. It's found in some of the most beautiful places in the world. It's not attainable by everyone. And it even gets you a little buzzed, which creates the feeling of "indulging and living life to the highest level." That's nice.

Our little wine industry has all the elements this consumer seeks. It's really a perfect match.

SECTION VI: THE IMPLEMENTATION OF WINE MARKETING

"The discovery of a wine is of greater moment than the discovery of a constellation. The universe is too full of stars."
— *Benjamin Franklin*

You Have to *Tell* a Story Before You Can *Sell* a Story

At this point, it begins to make sense that in the interest of time and capital, you can begin to explore some of the sexier tasks associated with wine, such as sourcing and creating, location, label design, digital media buildouts, etc. You know you have been chomping at the bit since the beginning, so you have the green light. Not to say you should put anything on firm ground, but it's safe to start exploring what can and cannot be accomplished and get some things in order. Again, quite a few books have been dedicated to that side of the business, so we'll stay focused on sales and marketing in this guide.

You have a Passion Statement to set the tone, you have dedicated time and resources to putting sales and marketing as a priority early in the process, you are committed to building and supporting a culture of excellence from top to bottom, you have a long-term strategy to either keep it boutique or scale it larger, and you have an excellent understanding of your customer. You have accomplished quite a bit, so be proud that you are already well ahead of most at this point.

Now is the time to start building out some of the details, starting with the Marketing Plan.

Again, for a few reasons, the methodology presented here is generally a simpler approach than the incredibly detailed versions I was asked to create in larger wineries. The first reason is to save you the extensive time and countless resources that are normally required to complete a comprehensive marketing plan. And more importantly, I truly don't think it is necessary. For some odd reason, many wine executives think a short, concise plan is not as complete, just based upon its size alone. However, I've led huge wine brands and I've led small wine brands. Both could be managed very effectively with brief, well-thought-out, crisp plans. Also, I find that the internal team as well as the outside partners appreciate the brevity. The goal is to execute and adjust along the way, not bury yourself and your team in planning.

Knowing Your Career Wine Marketing Professional

A brief note on getting to know your wine marketing professional and what to look for when hiring these extremely important roles.

The marketing role in a wine company requires a person to have a vast knowledge of the wine business, which only comes after many quality years of experience with other wine companies, making mistakes and succeeding. For this reason and this reason alone, you will need to hire a person who is deeply experienced in wine. Not just a highly-talented marketing person, regardless of how excellent they are, because there are just too many details to know that are specific to wine for an outsider to learn, but an excellent person who has immeasurable wine knowledge and excellent experience.

Outstanding wine marketing professionals are unique in nature, and the most talented ones are difficult to find. I believe the marketing role in a successful wine company is the hub of a wheel. Each spoke is a segment of your company - winemaking, sales, finance, etc. Marketing sets and drives most of the strategy and is involved in many of the initiatives so, naturally, they branch out into all parts of the business. If your marketing person is doing a great job, you will be able to see this easily because they will have many, various business units coming to them often. They should be one of the busiest people in the company because of this. Also, they must be generally experienced in all of these different parts of the wine business, unlike a winemaker or a human resources manager, for example, whose expertise is well-served in just one highly important business unit. The marketing person will make decisions that impact the entire company. Due to this, they should be highly skilled at handling multiple initiatives at once. In addition, they unquestionably should have the comfort and desire to get into the tactical details of an initiative, and exist way above the clouds in terms of strategizing. I've worked with too many great marketing talents that struggled with getting their hands dirty, so their effectiveness was always very limited.

Finally, and most importantly, I believe they need to have a great passion for wine. They will be the driving force behind communicating all messaging to the customer and the team, and a focal point for your business, which is based entirely on one product, wine. It needs real passion to translate the message to the costumer who is expecting it, or it will come across as "canned MBA speak". We've all seen the marketing of those brands.

Finding the right person for the position is not easy, and you should take time to really determine if a candidate is ideal. When you find your rock star, you will sense it. This person may benefit you greater than any other position in the company, so choose well.

The Brand Essence

This is an exciting step to take. Refill your large goblet of wine, sit in front of your laptop and, in a statement or two, define what your brand essence is all about. Basically, this is the heart and soul of the brand. It should read like a brand mantra or tagline, and in a simple way, encapsulate the spirit behind it.

For example, at Mumm Napa, we wanted to stay true to our sparkling wine roots that went well back to a luxury heritage in Champagne, but we acknowledged the fact that our customer base was all about being California-casual in a more lavish way. We envisioned the brand being like that of the feeling you get when visiting the Hamptons or Napa Valley, which is casual luxury. So, the brand essence was, "Mumm Napa is the quintessential embodiment of the casual luxury lifestyle." At Kendall-Jackson, the statement was more about the heritage of the founder, Jess Jackson, family and the amazing estate vineyards and wineries.

Before you get started, here are a few important things to keep in mind.

Firstly, be authentic and genuine. If you don't fully believe in the story, I can promise you no one else will. Or if you wax a little too poetic and blow it up to an unrealistic level, once again, it will come across as being unbelievable in the end. And this doesn't necessarily mean your brand personality has to be dreamy and wine geeky, with ancient vineyard blocks and black horses emerging from the morning fog. It can be an outstanding lifestyle brand about drinking rosé while at the beach with friends. Both can be believable, as long as it comes from an authentic place when you create it.

Secondly, don't overthink it. There are no right or wrong answers for a creative exercise. Trust me on this one because I've learned through experience, particularly this example. Many years and gray hairs ago, I was a young, cocky, newbie brand manager in the wine industry. While I was out in the market researching how our brands were being represented in wine shops, I came across a small wine brand. Just a few cases sitting alone in the corner, stacked three cases high with some bottles displayed on the top. Out of the thousands of bottles in this store, this one brand made me stop and stare. As I looked closer, the label had an image of a large foot. I'm sure my face scrunched up as I thought of how odd it was. You don't have to do extensive consumer research to know that bare feet on a wine bottle is not what most would consider an appetizing sales pitch for drinking wine. I recall thinking that this brand will be discontinued by my next visit, possibly sooner.

Fast-forward many years and even more gray hairs. As I write this, Barefoot Wine is the largest brand in the US marketplace and has been for quite a long time. The simplicity and accessibility of the brand essence seemed to resonate with millions of wine consumers and just kept growing in volume until it seemed as if it was everywhere. The original owner genuinely wanted to communicate a "hanging at the beach with friends" essence, and it worked. We all need these humbling moments to check ourselves and remember that creativity is subjective and only the customer will dictate what works and what does not.

So, take a break here. Use your goblet of grape juice as inspiration, channel your Hemingway, and put your pen to paper. This single statement will define your brand for everything it is, everything it does, and everything it becomes. No pressure. ;)

Your Brand Essence Statement

_____

_____

_____

_____

It's okay not to get it perfect the first try, or the first ten tries, because this will be a living-and-breathing statement that changes as you get to know the brand. This is especially true as you experience how the brand is perceived by your customers over time. Allow it to evolve, grow and mature. Each time, it only gets better.

The Brand Identity

Once you have established and feel good about your brand essence, next up is the brand identity. This is everything that identifies your brand, from its overall style, the logo, colors, website, social media channels, and so on. A good place to start is by asking yourself that if the brand were a person, how would you define them? Reserved and intelligent, adventurous and modern, traditional and stylish, casual and confident? Would they wear more muted colors like dark grays and blacks, or would they be showier like bright yellows and reds?

Referring back to a brand I managed, the identity was stylish, confident and smart, having a strong heritage to the legendary wine region of Sonoma County. It's colors were glossy gold and dark black with touches of bright red. Just from those simple two sentences, I'm sure you can formulate a picture in your mind about the brand's basic identity. So, once again, grab your goblet of grape juice, channel your J.K. Rowling, and put your pen to paper.

Your Brand Identity Statement

_____

_____

_____

_____

The Brand Positioning

Next up is the brand positioning, which is probably the most important statement to make in a quality marketing plan. It brings a sharp focus to the tactics of executing the plan itself. Primarily, it defines your wine's or service's unique value to your customers in relation to the marketplace and competition. For example, a classic luxury car company might use the following as its positioning statement: "The ultimate luxury driving machine for the discerning man, our car delivers a one-of-a-kind experience, immediately recognizable as powerful in feel, yet classic in expression." Once this is developed, you will use it to determine if your marketing strategy and tactics are supportive to the statement. If they are not, it's best to circle back and make the necessary adjustments to the plan.

To develop the ultimate statement, there are four intermediate steps to take that all feed the goal. These are micro-topics that create the platform for your positioning, so tackle these first because they will apply to the final brand positioning statement.

1. ***Target Audience***
   Start by briefly referring back to the section titled *Understating Your Customer*. That was a broad categorization of the overall wine-drinking population. Your goal here is to be just a bit more specific about the group of customers you want to reach with your marketing message. They are the people who are most likely to buy your wine or visit your winery or digital media property, and they usually are combined by some common characteristics like demographics and behaviors. For example, if your wine strategy is to focus on a $80 Russian River Valley-based Pinot Noir, your target audience would be a 'high net worth individual who currently buys and understands fine luxury wines, is 50-65 years of age, travels extensively, frequents 2-star-and-up restaurants, is married, and is an empty nester."

Your Target Audience

_____

_____

_____

_____

_____

2. Brand Category

The wine industry has a fairly standard way of defining brand categories, and it is based solely on retail price. I would suggest using this methodology because it is widely known throughout the business and will make it easier for people to understand where your brand is positioned.

Retail Price Range	Wine Category
$10	Value wine
$10 - $15	Popular Premium Wines
$14 - $20	Premium Wines
$20 - $30	Super-Premium Wines
$30 - $50	Ultra-Premium Wines
$50 - $100	Luxury Wines
$100+	Super-Luxury Wines

Your Brand Category

_____

_____

3. Customer Benefit

As an Italian from Jersey might say it, "Hey, what's in it for me?"

As you probably know, most people buy products with this very question in mind. The following very brief statement addresses that selfish, but very real, question. Referring back to our Russian River Pinot Noir example, it might be something like, "A distinguished and elegant Pinot Noir that few can obtain, it enhances fine cuisine and is always perfect for those very special moments with others."

Your Customer Benefit

_____

_____

_____

_____

_____

4. *Reason to Believe*

And finally, this is the no-BS declaration of how you will support the previous statements. Using our example, "From our renowned estate vineyards in the heart of one of the finest Pinot Noir regions in the world, each year this wine is made in small quantities to ensure its absolute perfection."

Your Reason To Believe

_____

_____

_____

_____

With those four intermediate steps in mind, grab your goblet and take some quality time here to really think about your positioning statement. When completed, it will be a profound and powerful messaging platform that will be a guiding light for years to come.

Your Brand Positioning Statement

_____

_____

_____

_____

The Wine Statement

The previous brand statements all cover many aspects of the brand, of course, and are fairly typical when creating a marketing plan. However, I believe having a separate statement for the wine is as important for many reasons. It gives style direction to your winemaking or sourcing team as they progress through making updated blends, crafting larger lots, refining smaller lots or combing the markets looking for new wines to purchases. Also, it puts the most artistically important part of your business, the wine, up on a pedestal giving it the hero status and the limelight it so deserves. Use as many descriptive words and adjectives as possible here. This is not intended to paint the winemaking or sourcing team into a box, but rather to guide them with a purpose as they put paint to canvas.

As an example, "The house style is one of prestige, intrigue and complexity, being boldly fruit forward, and masculine in style and depth. It can be enjoyed now or in the future with rewards in the development of exceptional aromas and flavors."

With this type of a statement, the sales and marketing team has an essence to work off of and re-communicate to the customer. And the winemaking team has a direction they rarely get. Also, I have found it opens an internal dialogue about the style of the wine that normally would not take place, so it can be an ever-evolving statement everyone can buy into and help develop together.

To understand this further, just look to the Champagne houses of France. With the very stringent government restrictions limiting crop sizes and having only three varietals to use in every blend, the house wine style is really the only unique difference that sets them apart from the other millions of bottles produced each year. Their wine statements are very well-defined, powerful platforms they will live and die by for hundreds of years.

Your Wine Statement

_____

_____

_____

_____

_____

The Sourcing Statement

Okay, I promise, this is the last statement you'll need to create, thus the reason I asked you to grab a *goblet* of wine at the onset. There's a lot to define here, but it's all for a good cause.

With wines that are not in a luxury category, the source of the fruit becomes less important. I can't tell you how many times I've written something to effect of, "The cool coastal regions of California bring the warmth of the sun during the day, and the coolness of the nearby Pacific Ocean at night, thus offering ideal conditions to grow this blah blah blah grape . . ." You get my point.

However, when it comes to luxury wine, the source of the fruit is of major significance. What we know about our customer is they like to be perceived as an "expert" on the wine they just found, so it's to your benefit to provide them with a few fine details. But, please, PLEASE don't do what many high-end wineries do, waxing poetic about the soil types or sun aspect of their vineyard. Most people won't read more than twenty percent of it, mostly because they just don't understand it.

I strongly suggest you make an honest descriptive statement about your estate vineyard or the source vineyard and location. Make it real. Make it personal. And most importantly, paint a picture knowing that most of the people who drink the wine will never visit the vineyards, winery or the place you purchased the juice.

This is your opportunity to bring them there, in the theater of their minds, through your eyes.

For example, "It's an early Sunday morning, and as I walk the rows, I can see the fog starting to lift. I feel the coolness of the ocean-scented air begin to dissipate as the sun peeks its crown over the tall eucalyptus trees that surround this tiny vineyard. The cool air is still, and the sound of a hawk hunting over our vineyard can be heard in the distance. The Cabernet grapes have been in love with this spot for years, giving us gifts of ripeness, density and elegance in the cellar every time, and this vintage is no exception."

Your Sourcing Statement

_____

_____

_____

_____

_____

Having Fun, But Not Too Much, With Packaging

Packaging is a very personal process and will be extremely specific to your brand, so what I can do is give you some solid, old-school guidance here. I've seen simple labels with standard bottles become great luxury brands; I've also seen iconic pieces of art with heavy bottles become great value brands. I've been designing labels and building brands most of my career, over 500+ brands at my last count. During these many years, I have developed a few clunkers and even more homeruns. I can safely say that there is no secret sauce because it is a highly subjective process with countless little details. So, to increase your chances of ending up with a label you're proud of, here are a few important things to keep in mind.

- *Get the best talent possible*
  If you don't have an experienced packaging expert on your team, hire a wine label designer (not just any designer) to help you with the process. You need someone who has experience in wine packaging since there are so many specifics that apply just to wine that other non-wine designers will miss. Most wine label designers are excellent and can interpret your

vision onto a canvas extremely well. They will be a sounding board, like a therapist, sifting through your thoughts to get your vision down on paper, and hopefully it will be a very cool-looking piece that you will love.

- *Use your marketing statements*
  Use the many statements you created earlier to drive your packaging process. Stay true to the brand essence and try to make the customer feel like they are getting something special. The wine consumer likes the perception that there was significant effort and craftsmanship involved. Make the final product feel like it's money well spent regardless of the price, and something they'd be proud to display for whomever they share it with.

- *It might be difficult but don't forget to KISS (keep it simple, ____)*
  Don't make this an all-consuming, complex process. I've seen how many people, especially new ventures, get very caught up in the creative process, going back and forth with the designer way too many times. In the end, it rarely helps to sell more wine. Also, the main risk is you can easily overthink the beauty and original purpose and take it right out of the process. Five to six total rounds should be your maximum. From there, you're just wasting time when you could be accomplishing other more important tasks that lie ahead.

- *Always know the costs and the production effort needed <u>while</u> designing, not after*
  Though you're designing for excellence, keep your costs in very close check, or at least on a spreadsheet in the vicinity as you make decisions. Never just ask for a larger bottle, etching, embossing, debossing, varnish, a hand-applied emblem, or whatever else you're dreaming up without knowing the production effort involved, the availability of the elements and the total cost. If your plan is to build a larger brand, this becomes even more important because the efficiency of creating many bottles at one time is paramount to the bottom line. Or, if you're plan is to offer a $100+ bottle of wine, this

doesn't give you the license to overpackage it. This can come across as trying too hard and can affect sales. Maybe you go with a heavier bottle, but the shipping weight becomes too heavy so less people click "purchase" at the end of their online shopping experience once they see the total cost to ship. There are many scenarios. Go big . . . that's fine. Or go simple . . . that's fine as well. But don't try too hard, and always take into consideration your costs and the production effort involved. The winery and finance teams will respect and possibly even love you for it someday . . . well, maybe not the finance team.

- ***Simplicity is the essence of wine***
  When designing a luxury product, as with wine, I strongly believe simplicity is key. This is not a 100% all-the-time statement because some crazy packaging concepts work great, but my experience has taught me that most wine customers resonate with simple and confident-looking brands that make a clear statement about who they are and what they offer. From a simple value Chardonnay to a luxury Cabernet Sauvignon, go take a look at the similarities on your local wine shop shelf. At the heart of most designs is simplicity. Also, do a Google search for luxury brands like Bentley, Dior, Breitling, French Laundry or the Ritz-Carlton. All the designs are elegant, simple, clean and they radiate confidence. Take a lesson from these brands that have defined their categories. They work.

- ***Plan way more time than you think***
  Though I've developed hundreds of wine brands, with an expert design team as well as on my own, almost every single time the process takes longer than I initially anticipate. Each time, I thought that since we had done it so many times with such experienced people, it would definitely be a shorter process *this time*. Then I was proven wrong as the process played out, as challenges were overcome, and as design elements were debated. After many years of design work, I've come to terms with the fact that it is a lengthy process involving great patience to get the desired results. Regardless of whether you're creating a brand from scratch or re-designing an existing brand, both scenarios take time to

develop. And if you have a team of people involved, add a few more weeks to the process because opinions play into the timing. I won't bore you with the details about why this is so, but take my word for it that the process can be 4-6 months from the start to when the label is on the bottle. So, plan extra time and be patient.

- ***You won't get it right the first time***
  Coming from a career wine brand developer and manager, I write this next part with a bit of disappointment, yet total acceptance. No matter how detailed you are or how hard you push the team, the final result will not be perfect… and that is okay. I keep telling myself this. The first time, what you're looking for is "as close to perfect as possible." The process is extremely complex with a multitude of designs, as well as printer and production hurdles to factor in. How they all come together is different for every design. There hasn't been a time when I held a bottle with a newly-designed label for the first time and didn't see three things I wanted to change right away. Maybe the black isn't black enough, or the embossing wasn't pronounced enough because the texture of the paper wasn't right, or the gold around the logo would have been better with a foil. You'll see it right away, and it's hard not to obsess over it.

Speaking from experience, and hours of transcendental meditation, follow this 9-step emergency plan when the moment arises of seeing a new design for the first time.

Step 1 : Breathe
Step 2 : Take note of what you would like to change
Step 3 : Breathe
Step 4 : Take a large sip of wine from your goblet
Step 5 : Breath
Step 6 : Understand that 99% of your customers will not notice what you're obsessing over, and what you designed is probably pretty excellent
Step 7 : Breathe

Step 8 : Start planning for the second round of label printing because this is where most labels become the *perfect design*
Step 9 : Finish the goblet and repour

The first time will not be perfect, no matter how hard you try. However, the second time most likely will be, so breathe easy, relax, drink more wine and everything will be just fine.

- ***Finally, don't look like "every vanity brand in Napa"***
  If you are one of the fortunate ones to have "made it" in other industries and have come to wine country looking to create a passion project, that is more than fine. There are many who have done this and several who have succeeded. But beware of what I call "the vanity packaging". You know, the one with the cool names like Titaniumus or Panpharmacon, and with blingy packaging that gives the impression that more money was spent on the bottle than in the bottle. I joke, of course… well, kind of, but I'm sure you understand my point.

  Take a close look at Silver Oak. It is a simple, glossy, flat label displaying a water tower design in a fairly inexpensive glass mold, and it has sold millions of cases as one of the original cult wines. William Selyem has an old school, funky, 70's design and is one of the most sought-after Pinot Noir brands in California. And don't get me started about the famous Burgundian winery Domaine de la Romanée-Conti because it takes a literary expert to navigate all the text on that label, and it sold for $15,000 per bottle.

  If your family name is important, then use it. If there is a rock on your property that you love, for whatever creepy reason, then use it. Be authentic, be real and be true to yourself. It will translate that way to your customers.

Designing packaging is very exciting, I get it. This is probably the most immediate enjoyment you'll be able to associate with and understand quickly, but (and I write this mostly to the wine biz outsiders with all intended respect) this doesn't make you a wine packaging expert. Understand your strengths and weaknesses. Hire a seasoned wine packaging professional to assist you in this process. Keep it simple, luxurious, clean and confident . . . and most of all, make it very personal to you and your brand. It's your baby, after all.

8 Imperatives of Pricing

Okay, now we're getting to the meat and potatoes of the process. This is one of the most difficult "throw spaghetti on the wall" practices there is because the price ranges in wine can be vast. However, there is somewhat of a science to it and it doesn't necessarily need to be an immensely extensive and expensive process. Attempting to use analytics only will only get you a pricing model that is not in touch with the street-level needs. Flying by the seat of your pants will only get you a pricing model that is either too low and loses money, or too high and loses customers. It is a process that requires some data, as well as an expert feel for the marketplace. This is a great place that your excellent marketing professional comes into play. Money well spent.

Instead of offering a complex, and fairly boring, spreadsheeted pricing model here, it is more important for you to have the philosophical rules that govern the process. Having these eight rules will save you not only time and higher costs, but more importantly, lost revenue.

1. *Check the competition*
   Start with researching brands that will be, or are currently, your direct competition. If you have a single vineyard Howell Mountain Cabernet Sauvignon, see what comparable brands are priced at and take the average. There is a good chance most of these competitors came to this price through trial and error, something you won't need to go through as you're benefitting from their experiences. This is your baseline price. For this example, let's assume the average of eight existing Howell Mountain Cab's come out to $110 per bottle.

2. *Factor in supply and demand*

What does your supply look like versus the demand for the region? Is it a few hundred cases or thousands of cases? Is the region known for small lots or can customers find it most anywhere? Scarcity or abundance of your product and category must be factored in. In Burgundy, most of the brands are from very small domaines, and since there is such high demand throughout the world, they can command a higher pricing model. In Russian River, though Pinot Noir has an exceedingly high prestige, it is in greater supply, so even the most luxury brands keep their pricing at comparatively middle to lower levels. There are always exceptions and they usually materialize from a history of either shortages or surpluses. Let's say your supply is small, that your strategy is to stay boutique, and the category is known for short supplies. It would be safe to increase your price to $150 per bottle.

3. *Always know your costs*

As a prudent winery or wine brand leader, you should absolutely know your COGS down to the bottle. This is more of a checkmark than a reason to price, meaning if your costs are high, it doesn't mean you should price your wine even higher. This is not a pricing strategy, it is a "I don't want to lose money" method, but most likely will take you out of the competitive and customer-acceptable range. If your costs are high, then focus on lowering them somehow, but pricing is not a lever to pull. If you use high costs as a pricing reasoning, you will only be faced with lowering your pricing later on as sales decline, and that only damages your brand. It's better to make far less on net margin and take the time to figure out production issues than to just immediately price your wine out of the category. Pricing is based upon the value-to-price perception in your customers' minds, not on your COGS.

But assuming your costs are in check, knowing the COGS per bottle empowers you with knowledge that good things are happening, like making a profit. And more importantly, if you are faced with the decision to discount at some point, for whatever reason, you won't bankrupt the business or damage the brand. Winery financials can be extremely complex and fluid, but they don't have to be rocket science. Just take the time to obtain an as-highly-accurate-as-possible COGS per bottle and know it well.

4. *Once you discount, you can't go back*
Regardless of your business strategy, one thing is very true . . . once you use discounting as a tool to drive increased sales, there is no going back.

If your plan is to have a larger, scalable business model, then naturally you will need to sell more cases to achieve these goals. The only places to sell more cases is in larger, off-premise retailers that have lots of wine on the shelves, or some digital properties that move a lot of wine. This also means heaps of competition sitting directly next to your brand. Also, depending upon what research study you're reading and whether you are selling online or off the retail shelf, at the point of purchase the final price is one of the top two reasons a customer ultimately buys a wine. So in these retail environments, the discounting game becomes a serious undertaking for all brands, even at the higher price points. I'm sure you have seen the level of discounting that takes place. Sometimes it looks like every brand has some price breaks almost all the time.

If you are new to the game, this next point might surprise you even more. Who do you think has to financially support these discounts? Since it appears that the discounted price is coming from the retailer, it must be them, right? Well, nothing could be further from the truth as it is *You* that pays for the majority of the discounting. All price breaks come from the brand owners. Very rarely does the retailer put skin in the game and support sales increases with discounts. If you want your price lower and more competitive on the shelf, then you will be paying for it.

Once you play the discounting game, you have now conditioned your customers to think they can buy your wine at a better price and you can never go back. They will buy it when it is on-sale, but they will buy a competitor's wine when yours is not on-sale. There is a very small group of über-loyalists who will buy a brand no matter the price point, but this is definitely the exception rather than the rule.

In a different scenario, let's say your business model is largely D2C, either online or in a winery hospitality setting. When wine companies discount here, it is not driven by competitive pressures because the customer is isolated in your digital or winery worlds. It is always driven by financial demands and objectives. For example, if you have strict sales goals in place, the team will want to use discounting of either the wine or shipping costs to attract more sales dollars. It is the classic marketing tactic. Discounting equals selling more. It becomes a need-to-have method to increase sales to the goals. And, as with the retail scenarios covered earlier, it is a tool that cannot be taken away once it is used. Your sales team might be compensated on sales goals, and without this lever to pull, they will not be particularly happy when it is no longer available.

Also, and interestingly enough, Amazon has affected how most everyone buys products online . . . basically with low or no shipping costs. And your customers will expect the same, especially as it applies to wine, because the shipping costs are usually high after factoring in heavier weight and special handling for the carrier who needs an adult signature. Once again, utilizing discounted shipping rates absolutely does increase check-out rates, and your online or club customer will expect them with every purchase from there on out.

Discounts are very common for value and luxury wine brands and work very well to drive sales. However, once you go there, again, *there is no going back.*

5. ***The ratings and awards matter, some of the time***
   This is a great debate in the winery industry.

First, let's make sure we're on the same page with terminology.

Ratings are the 100-point scales, or a methodology close to this, that most critics and wine publications use to convey the quality of the wine. Great wines are in the 90's. High 90's is excellent. 80's is okay at best. High 80's is even better for value wines. 70's indicates a wine of poor quality, or cooking use, whatever you like best. Candidly, the 70's never made total sense to me as isn't that a C grade? I don't know about you, but C was a pretty acceptable passing grade when I was in school.

Awards are a bit different, but just as easy to poke fun at. There is gold, silver, bronze, best of show, best of class, best of the universe, best on a Monday when the moon is full, best white, best kind-of-red, best with food, etc. They typically come from festival competitions that are put on around the country or at times when newspapers have strong wine sections, and they all have slightly different ways of calling out the top wines of that particular tasting.

At the sub-$50 price points, I truly believe the ratings don't matter to most everyone but the winery team and sometimes the retail buyer. The customer will buy your wine based upon a personal recommendation, the price point along with factoring in the discounted final price, and the label design. We've lulled ourselves to believe that these three factors mean less than a good rating or award. That is called eating our own dog food. If your wine business plan is to wade around in the sub-$50 pond, then spend your time with more important tasks like playing with the winery dog rather than chasing scores and awards.

At the over $50 price points, and especially the upper-luxury levels, like it or not, the handful of industry-respected critics remain one of the key factors of quality measurement and buyer perception, thus heavily influencing price. If you have an existing brand, I'm sure you've already submitted it for ratings. If you're just starting out, brace yourself because you will have to go there at some point. And in the luxury wine category, the cold fact is the ratings matter much more than they do at lower price points. This will be outlined further in the next section, Embracing the Critics. Focusing on the $50+ categories, generally speaking, anything between 96-99 points from the top publications should give you a nice increase in price. Anything between 90-95 should hold or justify your current price. Anything below 90 will make it hard to justify current pricing and could require an adjustment lower for the vintage. And finally, a 100-point score, especially if you get this for more than one vintage, will give you a cult-like status for that vintage only, not in perpetuity. 100-points are excellent, and they have a nice halo effect for the brand image, but they are usually as good as that particular vintage allocation is available.

Using our Howell Mountain Cabernet Sauvignon example, let's assume you received a somewhat disappointing 92-point score from The Wine Advocate, one of the top wine critic entities. This would bring you down from $150 to $130 per bottle. If you're new and have not received any scores just yet, then hold the price at $150.

6. ***Below $50 per bottle, going up in price is harder than going down***
Given the price sensitivity and impact to the customers' buying decision, as you can imagine, going up in price is almost never well-received and you will experience a dip in sales as a result. There are just too many options out there for your customer to easily jump into overnight. On the other hand, going down is fine as long as the price decrease is not so significant that it makes your customer wonder if there is something wrong. So, when pricing a new value wine, it is best to lean toward the upper limit of the category. This will give you a little room to lower the price if the situation calls for it.

7. ***Above $50 per bottle, going up in price is easier than going down***
Pricing a luxury wine brand is completely the opposite of pricing a lower-priced wine brand.

At the higher level, it's acceptable to increase pricing, but definitely not recommended to lower the price unless ratings put pressure on a particular vintage. The perception of a downward price is negative to the luxury consumer. It sends the message that something is possibly wrong, or that now many others can have something that they felt was special to them. Slight upward ticks in price show demand and give the feeling of scarcity. A slight price increase gives the perception that "I've got to have this wine.

I bet others are starting to find it and like it, too." I would even go so far as to advise that increasing the price each year, regardless of demand and what you have in inventory, will build on that velvet rope perception.

Wrapping up our Howell Mountain Cabernet Sauvignon example, your final bottle price is $130 this year. Your next vintage will increase to $135, and so on, as long as you see a sales increase as a result. If there is a stagnant sales year, then hold the price from the previous vintage, but don't go backwards if you can help it. As you do this, always keep a close eye on sales to see how the price increase was received. There will be some pushback, but you're looking for the majority answer, not the feedback of just a few.

8. *The promised land*
The most coveted and exceptional of all scenarios is that your own customer base buys you out of each vintage stock and you have a growing waiting list. If you have this scenario, congratulations! You're doing something great and your customers are rewarding you for it. This raises you above the marketplace and allows you to set your own standards to some degree. I will warn you, though - pigs get slaughtered in wine as they do in other businesses. Be proud and be humbled, but don't take advantage of the situation. If your wine is destined to rise to the price of rarified air, it should take time and rise in small increments to get there. Each uptick will test the viability of the demand. Enjoy the ride.

Starting with the competition to give you a baseline, then factoring in supply, scores, costs, and the knowledge that you can adjust pricing if you start off in a smart place, should give you good direction to price your wine accordingly. However, there is no grape-stained silver bullet here. Keep it on the conservative side. Take your ego and any cost issues in the winery or business out of the process. Once you start selling the wine, the customer will tell you how well you did; that, I promise.

Accepting, and Maybe Even Embracing, the Critics

This complex world can be narrowed down to two simple scenarios, regardless of your business model strategy.

The first scenario is you are planning on offering a non-luxury level wine, which is anything under $50 a bottle retail, and it could be either a boutique or scalable brand. If you read the pricing section earlier, you already have my take on this - I don't believe scores matter that much to your overall success. It is too much of a slippery slope to try and wait for a good score. Instead, I suggest using that time and those resources for engaging and strengthening the bond between you and your customer. If your wine, pricing and experience are great, your customer will follow you regardless of what score you get.

The second scenario is that you are planning a +$50 luxury-level wine, and this can be boutique or scalable models as well. Scores do matter more to higher educated wine consumers. Typically, this customer demographic is willing to pay more per bottle, thus their level of interest in wine makes it more likely they have visited the geekier wine publications and online wine sites. Inherent in this is exposure to the critics and their influence on the industry. They buy into this idea that these experts will know more since they taste so many wines, so scores matter to them, which means they have to matter to you as well to some degree, even though there are always exceptions to this rule and some successful luxury wine brands have bad critic scores. Nothing is easy, right?

There has been plenty written about the power and importance of wine critics, so we'll skip over the debate portion and get right to the how-to and with-whom portions. When it comes to the luxury wine category, you should only be associating with the top critics because they have most of the influence in luxury wine.

The list below is organized in order of impact to your brand. If they are not listed here but are highly influential, like Neal Martin of Vinous, it means they do not rate wines from United States and focus on other regions around the world.

As of 2020, this is a list of the most influential critics:
- Lisa Perrotti-Brown: The Wine Advocate
- James Suckling: JamesSuckling.com
- Antonio Galloni: Vinous.com
- Kim Marcus: Wine Spectator
- James Molesworth: Wine Spectator
- Jeb Dunnuck: jebdunnuck.com

If you're looking to export, the following would be important:
- Jancis Robinson (Europe): JancisRobinson.com
- Jeannie Cho Lee (Asia): JeannieChoLee.com
- Panel Tasting (Worldwide): Decanter

Next, engaging and having them rate your wine is not a particularly easy task. There are thousands of wine brands seeking their attention every day with only so much of them to go around. They are selective as to what wines they taste, and especially who they visit.

Inviting critics to your winery for a tasting is always paramount in importance to any other scenario. With a new wine release, I would only encourage a tasting *if they visited*. Send special invitations. Some will take the opportunity to visit, hear the story from the owner and winemaker, and taste the wines. Usually this makes the experience more intimate, and it can justify a higher score by the critic. It also gives you the opportunity to hear and experience their tasting methodology firsthand and absorb little nuggets of how to improve your score in the future. Finally, this establishes a great personal relationship with the critic that can and should be fostered over a long period of time.

If they do not or will not visit, I leave it up to you if the wine should be sent to them for tasting and rating. It's 50/50 for me, and this decision should be based on the critic's history of rating the region as well as any past experiences and ratings with your brand. For example, some critics think Pinot Noir is not desirable in the Los Carneros Napa/Sonoma region, so if you're a Pinot house in this area, I would steer clear. Ultimately, being patient to present your wine in a personal format on your turf is always the optimal situation.

With luxury wine, like it or not, the scores matter quite a bit. Social media is so prevalent that your customer can easily look up your past scores within seconds. Fair or not, it's the Yelp reality we live in today. So, with this in mind, here's a quick guide on the average score levels and the impact they provide:

- *Below 90 pts*
  Your wine program needs to be reevaluated to decide if this is a one-off anomaly bad score, or something more that requires a paradigm shift in style and quality. Basically, a little soul-searching with you and the team is needed.

- *91-95 pts*
  Not bad. This doesn't help pricing or demand much, but it can substantiate your current place in the market. This rating should send the message internally that the wine could easily slip into the high 80's, or just as easily, with a solid plan and some execution, rise up to 95+.

- *96-99 pts*
  This is a solid score and something to be very proud of. This could justify a price increase, thus creating a higher product demand and facilitating a rise in stature for the brand essence. If you find yourself here, this is a safe range you and the team should be very proud of.

- *100 pts*
  These scores don't happen often. Be careful what you ask for, though, because staying here is harder than getting here. (Ask any 3-star Michelin chef.) Soak in and savor the moment, too, because the halo effect of the score can be used for years well after the vintage. It amplifies the entire brand, as well as any other projects like secondary brands. The team, the estate, the vineyards, and the winery now become as renowned as the wine itself.

Critics . . . love 'em, hate 'em . . . the debate goes on. Regardless of what side of the table you are on, at the luxury wine level, the critics have great influence, so play the game but play it on your terms. It can make a significant difference. However, if you don't have to or want to *go there*, then by all means DON'T! You can still be highly successful, as many of the non-rated or poorly rated wine brands that have come before you have proven.

But what if you got a 100-point score? Hmmm.

2 Purposes & 10 Imperatives of Social Media

Regardless of customer demographics, without question, social media is a very important marketing communication tool today. Though there are many apps that provide wine information, here we will focus on the platforms you will be using to engage with your customers directly. We will be covering the usage of an email mailing list later in the sales sections.

In general, the wine customer uses Facebook and Instagram almost exclusively. I'm sure this will change as technology evolves, but at the time of this writing these two platforms rule and account for nearly all of the social media customer engagements. And I'm sure people can make an argument for other social media platforms like YouTube, Pinterest, Twitter, Snapchat or LinkedIn. But I truly believe that a focused effort is much more effective and spreading your messaging across more than two social media platforms is a waste of valuable time. So focus on the most used by almost all wine customers and your results will be better.

Wine social media serves two very distinct purposes:

1. *To reinforce brand essence and messaging*
   Here is an opportunity to continue to build the perception of your brand and the connection to your customers. If you're a luxury brand, post images and short videos of the artisanal sides of harvest, customer visits, your wine in interesting places, the cellar team racking barrels, bud break in the estate vineyard, and your

team designing new labels. Tell stories, educate and, above all, inspire.

If you are in a non-luxury category and more of a lifestyle brand, then post images and short videos of the wine in the settings that match your brand essence. If you have an awesome rosé, then maybe images at the beach in a social setting would be more fitting. Or if you're brand is about food and wine pairing, post photos of the wine next to food, possibly incorporating recipes for customers to try at home. Not perfect examples, but I think you get my point.

2. *To communicate important events*
Secondly, social media is an ideal place to communicate events, partnerships, winemaker dinners, market visits, new releases and anything else that is informational. Facebook has a built-in platform for scheduling and communicating special events, so this may work better overall. However, Instagram is more like a billboard and should be used as well.

I can imagine there could be a whole book written on wine social media alone. However, the purpose here is to acknowledge that the wine industry has a million moving parts, thus we need things distilled down to sound bites and simple tasks. And in the interest of those sound bites and simple tasks, here are the top 10 imperatives you should know when engaging your customer with wine social media.

1. *One person drives the bus*
It's important that you have one person on your marketing team manage everything, whether it's public-facing social media posts or inbound communications. This allows you to have a trained person monitoring the messaging and imagery of your brand perfectly, which includes answering customers, timing and frequency of posts.

2. *Have a general plan and guidelines*
Social media is not rocket science and definitely not as complex as making fine wine. Having a general plan for what to post and where, how frequently, and how it should look is a very good idea.

For example, a simple plan could be to post on both platforms at least twice per week, use stories on a more in-the-moment basis, have bright colors and inspirational images, and post at least one video per month.

3. *Have authenticity in everything you post*
   I put this one toward the top because I believe it is one of the most important aspects of social media. If your posts look too perfectly curated, with smiling people at a picnic table sharing wine, and I would mostly likely see something like it on a dentist's office wall, then that is what everyone will see as well. Boring, common and so very inauthentic. All of your posts should have a high level of authenticity. Keep it very real. The customer will appreciate the genuineness. You can still have fun and post cool people at a picnic, but make sure it is real people maybe at one of your events, not stock images you buy online.

4. *Always stay within the brand identity and essence*
   Regardless of your wine price points and category, everything that is posted should be well thought out and perfectly in line with the brand identity. If you are a lifestyle brand that is associated with handcrafted wines, your posts should have an artisan, handcrafted appeal. You wouldn't see a luxury Champagne brand posting images of forklifts in their caves, but you would see a Chef De Cave winemaker artfully holding up a bottle and inspecting it as it ages en tirage.

5. *Get your customers to participate*
   This is also an opportunity to ask your customers to share their experiences as they visit the winery or sharing moments at home as they enjoy your wine. You can do this via a post or using marketing materials in your winery, inviting your customers to engage with your channels. Encourage your customers to tag your brand when posting. They will post inspiring pictures from around the world in amazing locations with your wine at the center. This is also a way to keep your brand at the top of your customers' minds and create a loyal personal relationship with them.

6. *Don't oversell*

Social media platforms are not proven ways for effective wine selling just yet. Overall, these platforms are there to help build your brand essence perception and communicate cool events, which will enhance and build a real connection between you and your customers for purchases later. But overselling with discounts and promotions will get very old to the average wine consumer and will result in "unfollows" and thumbs-down's very quickly.

This is not to say you shouldn't promote special deals. If so, use Stories as they will be seen more frequently, can have web visits incorporated, and are used more generally for this type of activity. If your customer is not interested, it goes away within a day. Instagram is the best platform for this type of activity because customers can swipe up to quickly see the specifics of your deal.

7. ***Don't forget, it is a two-way street***
There is a school of thought that suggests you should visit other like-minded social media channels and tag them to generate traffic back to your sites. This is true; however, what is also true is that it takes a very long time and lots of tagging. Just a little doesn't move the needle. So, unless you're paying your teenage daughter to do this 24-hours a day, for the time being, I would suggest focusing your teams limited time and resources on your channels alone and don't actively tag other sites.

And, most importantly, have your lead social media person reply to comments and messages at least once per day. This can easily be done using Notifications, which will notify them when a message comes in. Your customers love to be heard and acknowledged. If they take the time to post "Love your wine!", replying with a big thank you and inviting them back to the winery is a great way into their little social media hearts.

8. ***If you are asked to promote a post, put the phone down and walk away***
Facebook is an enormous company because they are excellent at encouraging promotions. If you are unfamiliar with this, I'll explain further. If you have 1,000 followers, when you post, it does not go out to all 1,000 followers. It is only exposed to about

30-50% of that group. If the post is highly "liked", Facebook will share it a little further based upon its successful engagement. However, if you want it to be exposed to all of your followers, and possible beyond your universe and get non-following wine geeks engaged, Facebook will gladly take as much money as you would like to spend to blast it out further. And when that little notice of, "Hey, would you like your post to be seen by thousands of wine customers" pops up, it's hard not to get excited and click on the Promote button.

However, I'm going to ask you that when you are in that moment to put the phone down and slowly walk away. It is not worth the time and money to try and get your post wedged into the massive amount of other posts and advertisings that occur on one person's feed. It's like trying to out-advertise Toyota. It requires very high frequency and thus very large budgets to make an impact, and for little gain other than a "like" or a "click" ~1% of the time. Yay.

For those of you who have a D2C online business model only, you can use the social media promotion machine to your advantage, but I strongly suggest planning very large budgets. I have been involved with some of the largest online wine companies, and they spend millions of dollars to gain the same number of customers a high-traffic winery might get. I suggest using those millions on other more creative ideas to attract and engage prospective customers, like pop-up dinners in local restaurants, or home wine tastings. It will get you further in the end.

9. *Use video where you can*
   It is a great idea to use video as much as possible. Data shows that people typically only view up to 15-20 seconds of a post, so keep it short, concise and meaningful. Choose videos that will educate, create inspirational moments, and again, bring your customer closer to your brand. Use your smartphone along with a steady cam attachment like a gimbal or a tripod. That's all you need. Video is an amazing way to bring to life the messaging and personality of your outstanding team and brand.

10. *Keep it up to date, or the opposite effect occurs*

A way to have social media backfire is to either post something offensive, which I'm going to assume will not happen, but also to have an infrequent or dead channel. When a prospective or excited new customer checks out your brand online only to find the last post was five months ago, it sends only one message - that the company is out of touch and not buttoned-up, so the wine and experience will probably follow suit as well. Keep your sites up to date even if the posts are not stellar, or the opposite effect will take place.

Social media should not be a time-consuming initiative; rather, it should be an exciting, informative, educational and inspirational way to engage with your customers and business partners. So keep it light, keep it on brand, and keep it real.

An Inspiring Website

This is your best opportunity to inspire and sell wine.

Here is the most important platform to communicate the marketing essence and identity you have spent plenty of quality time putting together, and to hopefully sell a few thousand cases of wine. This is your place to tell your story and inspire. It is a complete reflection of your business and your wine, so take great care in how you represent the brand. Use some of those awesome iPhone images you took, and even post a few short videos. Go ahead and be a little cheesy, it's okay to do that here. Be authentic to the brand story and keep your website extremely simple and intuitive. As your customers are on the site, their tolerance to find what they need is very low. They want to learn but not be inundated with information. And most importantly, don't forget that this is wine, so they want to be inspired and impressed. Think of it as a billboard. Once you launch the site and look at the analytics, you will see that most visitors spend very little time on each page, so do yourself a favor and design with simplicity and ease in mind.

Even with the massive impact that social media has on the way we engage each other in the world, the website has remained the most important communication platform for your wine company and brands. It is where the action happens in terms of more information on your branding, wines, hospitality services, and the best place to purchase wine. Consider where you purchase most anything online today. From a Ticonderoga pencil to a Porsche, just about everything is accessible with a credit card, Internet connection and a shipping address. It is actually quite a brilliant and very powerful time for wine companies to express their brand images and, most importantly, sell more wine.

Many years ago, while I was managing a large wine brand, we had an initiative to rebuild the website. We hired an outside firm, took six months to plan the layout and flow, spent well over $50k, and created a very nice website. Well, times have changed considerably since then. With the sophistication of many website creation and hosting companies, some focused entirely on wine, it has never been easier, faster and more self-sufficient to take on the project inhouse. Given that this guide is intended for start-up to mid-sized wineries, the few solutions below are more turnkey and can be initiated without many resources or breaking the bank account.

There are simply two paths to consider. First and foremost, you will need to decide if you will be selling wine online, either as a marketplace or as a club, or merely offering an information-only site.

1. *Information-only*
   This makes your life much easier, for sure. There are three leading website creation and hosting companies in the marketplace today, SquareSpace, Wix and WordPress. They are all extremely intuitive and inexpensive platforms that make it almost too easy to build and launch a truly outstanding website. Deciding upon which to use is a personal preference because they all have slight differences. If you're not that handy with technology, there are plenty of consultants that are experts with these platforms, and for a few extra dollars, they can bang out a great site for you.

2. *Selling wine*

This becomes a bit more complex, but again, given the explosion of D2C wine sales in the US, a few offerings have also emerged as leading providers. For example, VinoShipper, WineDirect, Commerce 7 and VineSpring. They all have very unique capabilities and solutions, so who you select will depend upon the type of business strategy you have, your situation, and the licenses you hold. It will take a little time to go through and understand all of these solutions, so plan ahead and be diligent. Once you're locked into a solution, it is not terribly easy to change later on. What? Did I hear that you have a marketing expert in-house that has experience in these areas? Interesting.

Later on, in the sales sections, we will cover wine clubs in more detail. So select a great platform that is easy for you to manage and use. And, most importantly, be creative and inspire . . . this is your moment to shine.

Developing Partnerships to Create Experiences

Here's a very interesting topic of differentiation, that being brand associations. This is something the majority of wine companies do not engage in. Most hope their customers find their wine by either visiting the winery in person or by visiting their local wine shop, where their brand is presented amongst the other thousand or so on the shelf. This doesn't feel very proactive in providing a more complete luxury experience, does it?

Let's go back to our customer profile and review what resonates with them. Wine is considered a luxury lifestyle that incorporates many elements like the beauty of wine country, excellent wineries, outstanding food, and travel. They are not attracted to a bunch of products in isolation tossed at them randomly. They are attracted to exclusive experiences that can surprise and impress them.

Here is where you get to be very creative by bundling or associating your brand with other like-minded brands in unique ways to stand out from the crowd while targeting your key demographics. Start by writing down all the categories your brand associates with, such as golf, resorts, top restaurants, architecture, music, art, etc. In each category, identify the top players. Then formulate a plan and have your marketing team contact each target brand with a well-thought-out partnership offer. Make sure to provide wine, as few people will turn this down, and a piece of follow-up material. If you get a hook, most likely you will need to visit the location or try the product to solidify the partnership and make sure it's executed in the way you've envisioned it. Here are a few examples:

1. *Online associations*
   When you send out a wine shipment to a customer's home, or even in your tasting room as they check out, provide an excellent partnership offer. If you have a big red wine, partner with an online meat company and offer a special, discounted, customer-only price to pair your wine with their meats. Also, if you have an excellent white wine, build a partnership with an online seafood company. This can work with glassware, spices, candles, you name it. It's unique and impactful, and you're providing more of a complete lifestyle, not just wine.

2. *In-home tastings*
   Many high-income people love to throw in-home, catered parties. Partner with luxury catering companies in all the top cities. Offer them outstanding pricing and any support needed to convey the style and essence of the wine. Even offer them a representative to pour the wine. Have someone there to present the wine and do a food pairing. If it's legal, you can even take orders. That's right, you're always selling.

3. *Local club partnerships*
   Partner with the local clubs like fitness centers, private golf, yacht or members-only social clubs in sensibly selected cities. Offer their members a private tasting event, which could include bringing in local chefs for a pairing, or an exclusively curated dinner event.

4. *Local hotel partnerships*

Target the top-20 small boutique hotel properties in the country. Visit them and offer a great discount to provide them with your wine they can pour as the customer's reception or dinner gift. Give their customers an exclusive deal to visit the winery, offer to purchase off the website, or a priority spot on the waiting list.

5. *And even clothing shop partnerships*
   In each city, there are very high-end clothing shops, usually custom clothing, that cater to the über-wealthy. Offer these shops a free case per month to pour in their shop as a gift to their customers. This works particularly well if you are offering bubbles. Even extend a private invitation to visit the winery and receive a priority spot on the waiting list.

You can reciprocate and benefit further from all of these ideas by using your mailing list to send out exclusive deals for your partners. For example, golf for four at the private Cypress Point Golf Club in Pebble Beach at an exclusive discount.

These are just a few ideas to get the creative juices flowing, and depending on your brand essence and personality, I'm sure you can come up with a few others that cater to your brand more specifically. Very few wine companies do this, and it could be your greatest source of growth and differentiation.

How Important is Public Relations?

Unless you have significant capital to spend, I would recommend not investing in public relations, at least for the first 5-8 years as the brand plan unfolds and you have a better understanding of where your business is developing. For a boutique luxury brand strategy, it's nice to get a mention in the Robb Report or any other luxury magazine, but I question the actual impact it has, especially after factoring in the costs of the firm you hire. It is very possible to accomplish what you need with direct customer relationships, partnerships and social media in today's world.

If the goal is to scale larger, then again I would suggest holding off on hiring a public relations expert until you develop the business. Once the brand has some volume, cash flow and a national footprint to it, then it might be a good initiative to hire someone who specializes in wine and drive the message home in a targeted way as the brand reaches out to a larger audience. A few mentions of the wine in the right places could support the growth and encourage new retail placements. I've seen PR work excellently for very large wine companies, but I have always questioned its value to a mid or small-sized wine brand. This is not a reflection on the talent and capabilities of the public relation teams themselves, this is only an observation of the true value impacting the brand in a meaningful way that drives sales.

If having the additional financial resources is not an option, be safe in knowing it won't make or break your wine venture, so you should use your money and team efforts in different, more creative ways.

SECTION VII: THE EXECUTION OF WINE SALES

*"Quickly, bring me a beaker of wine,
so that I may wet my mind and say
something clever."*
- Aristophanes

Own Your Fate, or Someone Else Will

Up to this point, outside of the "simple" tasks of crafting a world-class wine, you've done a lot of hard work defining the marketing of your brand. Consider yourself way ahead of the game and positioned for great success already. Some marketed initiatives feel like sales initiatives because the two do overlap quite a bit in the wine industry. When starting a new venture, marketing strategy will take up a considerable amount of time. However, when that exercise is fully developed, the pendulum swings quickly to the sales side and stays there for most of the time thereafter.

If this guide has advised on anything, it is that sales is the most important part of the company and can ultimately decide your fate between monumental success or having a lot of great wine to drink at home by yourself. I strongly believe that sales are 100% within your control, and the quality of the team and excellence of the execution will determine the height of that success.

The US wine market is massive and growing consistently, even during economic downturns. It is fairly new as compared to other world markets, and extremely healthy. We have enjoyed a long period of great prosperity in wine, so there is plenty to go around for all wine companies. All you need is your slice of the pie. This also means there is considerable competition with a lot of wine out on the market. This is your chance to put the marketing plan into action and execute at the highest level possible. Make your wine company a sales-driven culture and your slice of the pie will be big, and damn tasty.

The Wine Sales Channels Defined

There are many marketplaces, or sales channels, through which to sell wine. They require strategy, expertise, financial support, and time to manage. All have their pros and cons. As with crafting the wine, each needs to have a full commitment and myopic focus by management and the team or it will end up as a sub-par effort that never reaches any of the goals. When it comes to sales, this could be disastrous. So, deciding upon which ponds to fish in and staying committed to them is a smart way to approach your sales strategy.

Let's start by defining the four major wine marketplace segments.

1. *Direct-2-Consumer (D2C)*
    This is selling directly to your customer, either by shipping to their home/office or hosting them at a winery or tasting room. It's a fairly new and booming segment of the marketplace. For many years after prohibition, and until very recently, this was nearly an impossibility. With the help of a Supreme Court decision, most states have changed their archaic laws and are now allowing shipments of wine directly to the customer.

2. **3-Tier (Wholesale)**
    This is selling to the customer via a retailer, which in most states requires you to sell via the 3-tier system. This involves a very large, complex and established wholesale network of companies in each state.

3. **Direct-2-Business (D2B)**
    This is selling wine to other wine companies. There is a whole, very large business within the wine business that pertains to wine companies selling to wine companies. This will not be covered in this guide as it is a very unique, niche segment.

4. **Export**
    As it suggests, this is selling wine outside your own borders, typically through broker and distribution partners who have very specific knowledge and contacts of those countries.

In terms of where you should be spending most of your time and investment, it largely depends on your long-term strategy of either boutique or scalable, as well as your companies' structure of either being virtual or having a physical location like a winery or tasting room. D2C is clearly the most profitable and desirable because you have direct access to your customers; however, it is very costly to attract and engage new prospects. On the other hand, 3-tier has the most scalability for growth but is far less profitable. It is highly complex and competitive since most of the largest wine companies dominate in these marketplaces.

If you're still not sure, as you read further, the sales channels that are the most ideal for your company will become clear. It might be very narrow, like D2C online sales only, or very broad and cover a few different channels simultaneously. Regardless, always make the decision based upon what you can point a large part of your talent and resources toward. Spreading out too thin and dipping into all of the channels is a recipe for failure 100% of the time, so be selective.

Knowing the Career Wine Sales Professional

A brief note on getting to know your wine sales professional and what to look for when hiring these important roles.

I've mentioned it numerous times, so one more time won't hurt anyone - the majority of wine companies don't emphasize hiring quality salespeople enough, particularly in the start-up phase of development. It's common to employ an amazing, experienced winemaking viticulture and cellar team. However, if no one sells the wine at a high-level of expertise, then no one drinks the wine in the way and places you have envisioned.

The sales role in a wine company requires a person to have a vast knowledge of the wine business, which only comes with many quality years of experience making mistakes and succeeding in other wine companies. For this reason and this reason alone, you will need to hire a person who is deeply experienced in wine. Not a highly-talented salesperson, regardless of how excellent they are, because there are just too many details to know that are specific to wine for an outsider to learn. Using top restaurants as a comparison, think back to a time when you've been fortunate enough to eat in an outstanding restaurant with some of the finest food you've ever tasted. Usually the wait staff is exemplary in how they present and serve the food. Watch them because it's like a well-scripted, choreographed dance unfolding. At the 3-star Michelin level, they are required to be immensely knowledgeable and clearly experienced. It enhances the food, the restaurant, and the brand, but most importantly, it presents their product in the way they intended it to be enjoyed.

Next, they need to have a hunter's mentality. They are your warriors and selling wine is a very difficult, daily grind, I assure you. The finest sales representatives that have worked for me - and there have been many - are tremendously resilient and start each day with an outstanding drive and renewal of purpose. They are always hungry, and not just for commissions - they have a need to win that supersedes all other reasons to sell wine. Driving this is a healthy ego and the pretty constant desire to be heard and validated. They will be your most opinionated people on the team, and they will also generate the most revenue.

Finally, as with your marketing team, they need to have a passion and knowledge of wine. Their wine expertise can be just specific to the category your brand is in; however, their passion for wine is essential. It continues to create the culture you want to convey, internally and externally. They need to be vetted, trained, empowered, and given the tools needed as with any other part of your wine venture. And they need to embody excellence at every moment and in everything they do. Without it, you've compromised your investment, and your dream, severely.

For the Love of Direct-to-Consumer (D2C)

At the time of this writing, D2C wine is a $3 billion, six million-case category accounting for 10% of the overall US wine sales in 2019. This segment has an average price per bottle of $65 and it's growing at +9% annually. It boasts by far the highest margins, and more importantly, you can own the customer experience from top to bottom. For an industry that struggles to innovate and change, this is truly an opportunity for all who dare to make a career in the wine business. For any winery, this is the key to great success. And these outstanding data figures do not include all D2C because the tasting room and winery experiences are accounted for separately. So, needless to say, there are many reasons to make this segment one of your key focuses.

As it pertains to your business model of either keeping it boutique or scaling it larger, you would approach D2C differently. For a boutique strategy, this should be between 70%-100% of your focus and allocation of your most important resources. Your business model is just not ideal for the 3-tier segments. I've directly experienced how some small wineries attempt to increase revenue by launching out into the 3-tier stratosphere only to struggle given the extremely high costs and longer than expected lead times to build the business. This stretches the already limited resources these wine companies possess and takes focus away from the more profitable segments of the business.

For a scalable business model, the equations flip to 0%-30% focus of your resources allocated to D2C. When you're going to war and building larger brands, it requires the whole army to be in the struggle of the 3-tier system to win. I've also directly experienced how large wineries struggle to offer a more-than-normal visitation experience. However, if you have a winery with hospitality capabilities, then offering a D2C experience so that those new customers who love your brand from afar will want to get up close is very important. Beringer, Mondavi, and many others have some of the greatest hospitality experiences in wine country. It reinforces and strengthens that bond between your brand and your most loyal customer. However, beware that it doesn't take up more resources than the revenue it generates.

The 3-tier beast needs to be fed constantly. The winery traffic will dry up if neglected.

Breaking down wine D2C further, there are three major segments:
1. Tasting room hospitality
2. Wine club
3. Email and website

We will go into the details of each later in this section.

Go Beyond the Expected

If I visited ten wineries for tastings, I would get ten very similar experiences. This is an issue.

After all the dreaming, the investment, the years overcoming challenges to get to this point, in all actuality, you are offering an experience that is similar to most other wineries. Come in, sit down, cool place, nice people, good wine, buy a little, done. Apple inspired the world with their very simple, yet legendary marketing campaign called "Think Different." Pure genius. No one I'm aware of has revolutionized the world by saying "do the same as the guy down the street", but this is what you see down almost every wine trail. I still love the experience of an incredible winery, but I'm constantly amazed that many of these remain relatively the same to each other.

If you intend to use the rest of this guide as a supplement to Ambien in an effort to sleep at night, I get it, because it can be a little detailed about the specifics of the wine business. But I would say this is the most important piece to read and embrace, so grab a cappuccino and power through.

To be the best, you have to be unique and transcendent by curating a memorable and priceless experience for your customer. They need to walk away believing your experience was the finest they have ever seen. Your customer demographic demands it, your wine demands it, your commitment to excellence demands it, and your dream demands it. Use imagination and creativity to ultimately connect at a much deeper level. Don't just keep up with the Jones's down the road. You are in the wine business to create the greatest customer experience of all time. This is where you can set yourself apart from the pack.

You should always have the finest-trained, most knowledgeable, top employees as customer service representatives, and they should offer concierge-type services for any touch point needed. Your representative should know the name and contact information of all their top customers, and if they can establish a more personal relationship with them through occasional calls or via email, all the better. Ask them to go above and beyond. They should be perceived as being helpful for any request, be it wine, wine pairings, restaurants, hotels, renting a Harley in the area, or finding wine near their home. Anything is game. Your representative should be perceived as the customer's own personal wine concierge service.

Previously, we covered the pursuit of excellence in detail, and it applies here more than anywhere else. This is where all the pieces, the talent, the experience, the 10,000 hours of hard work, the countless dollars and dreams all come down to showtime. This is not just about getting customers to taste your wine and sign up for your club or wait list. This is an opportunity of much greater magnitude. Refer back to your knowledge of the wine customer. It's all about creating an unforgettable priceless experience . . . something they can talk about later and share with their inner circle, and something they want to be a part of long afterwards. It has to be about the magic and artistry of the moment in its totality. This means stepping back and questioning everything in this pursuit of excellence, and most importantly, from the perspective of your coveted wine customer.

Curate an experience that goes beyond the expected.

7 Imperatives of a Memorable Hospitality Experience

This is the most impactful, personal and principal customer interaction there is. Each situation can be unique depending upon brand essence, price points, business goals, etc. However, one thing holds true for them all, that if a customer makes the decision and takes the initiative to visit your "home", it requires your finest effort each and every time.

These seven imperatives, or suggestions really, are intended to be unique compared to many of the other hospitality experiences. The goal is to be memorable, to personalize your experience, be creative and take the liberty to add touches of your brilliance to the suggestions below.

1. *Start before they get to you*
   Make your first impression well before they walk through the door. Set up a partnership with a nearby four-star hotel or car rental company that offers a prestige line of cars. Offer a discount and an in-room gift with a personalized note welcoming them to wine country. Offer the concierge services of one of your representatives if the customer needs assistance booking restaurants or other winery visits. Also, welcome them to wine country personally by sending an email the day before. Of the three wineries they think of visiting the next day, who has made the best first impression? You have.

2. *Set the tone the second their foot hits the ground outside*
   From the moment they enter the winery or tasting room, all expectations and senses are on high alert with eager anticipation. Most wineries have the sight-thing down pretty well, but what about sound, smell and touch? Refer back to the last time you stayed at a luxury hotel. Most likely, the moment you entered the lobby you heard subtle music that aligned with the brand image, and there was a bright floral arrangement or an awe-inspiring piece of artwork in the entranceway.

   In the parking lot, which should be perfectly maintained because it's the first thing your customer sees, have music playing to create an atmosphere. Nothing appeals to the emotional core of a person more than music. Plant aromatic and colorful flowers in the surrounding area to inspire your customer with amazing smells and shades of beautiful color as they walk in. Continue the music as they walk through the front door. Dare to create Disney magic and moments of wonderment.

Well before they are greeted by your expert team, you've had multiple touch points setting up the greatest wine experience they will ever have.

3. ***The third impression is the personal greeting***
   There is nothing more important than greeting your customer with your undivided attention, a smile, and even better, a name recognition. If there is a dedicated greeter or a team member who will be hosting them for the day, they should be at the front door waiting to greet them by name, welcome them, set the agenda for the visit, and maybe even offer a glass of white wine. This should happen immediately and perfectly. First impressions always impact the rest of the experience. Think back at a restaurant that had a bad host, but great food . . . not perfect.

   Here is where you can influence another sense, that of touch, with a warm towelette. It sounds cheesy, but elite airlines offer them, and I've never seen anyone turn it down. It feels thoughtful, especially to luxury customers who stereotypically have a tendency to be more germophobic than most. I'm speaking from great experience.

   Again, these suggestions are not to be taken as the gold standard for everyone. These are ideas to expand your mind well beyond what one thinks as *a normal experience*, and to push the limits of being one of the best.

4. ***Make it a complete sensory-based tasting to remember***
   This is another perfect opportunity to differentiate your brand from all the others. The setting for the tasting should be truly excellent, whether in the winery, on the deck, or in whatever beautiful situation you envision. I strongly suggest a sit-down

tasting, which is more inviting and allows for an added relaxed atmosphere. Have everything perfectly placed and prepared *before* the guests arrive. Have subtle music playing in the background and make sure the setting is visually stunning as you continue to appeal to every sense.

Once your guests sit down in super comfortable chairs, your team should ask as many questions as possible to get them engaged and talking. After a few moments have passed and the ice has been broken, your hospitality expert should wax poetic about the owner, the winery, the estate vineyards, the winemaker (your rock star,) the house style, and the wines they are about to taste. This must be semi-scripted with each person's personality peppered in, concise and well-practiced while still feeling personal and off-the-cuff. Visual aids can help to bring the story to life. Overall, the message needs to be conveyed similarly and consistently from all your team members, with little variation. I strongly suggest you practice with your team often. I've experienced many wine professionals who will claim they can deliver a wine speech but less than 30% are excellent when it comes to primetime - understandably, because it is a difficult task for anyone to speak in front of others. So this cannot be left to chance and the only way to a higher level of excellence is through practice.

There is no glory in practice, but without practice, there is no glory at all.

Here's a unique idea. Regardless of how well-educated a wine consumer is, most still struggle with understanding all the subtle nuances in aromas and flavors. As experts, we know this all too well. We all listen to self-proclaimed cork-dorks go on and on about their wine collections and travels, but how many really know the complete tasting nuances and ways to accurately describe wine? I don't believe many do, because they just have not had the exposure to great wine and winemakers like we have for many years. Therefore, when your team talks about "aromas of dry leaves with hints of ripe cherries, blah, blah, blah…", know that most in the room have no idea what they are talking about, though the customers will nod their heads giving the illusion of understanding. Have the team ask qualifying questions along the way to make sure everyone is on the same page.

Another unique idea is keeping a sensory kit in the tasting room, like cool artisan jam jars with aromas and flavors in them. Before they dive into the tasting, go through the many characteristics found in your wines using the kit, like cherry, white pepper, green apple, mushroom, and so on. I know this may sound unorthodox but be confident in the fact that most everyone thinks they know these aromas, but they don't. Accentuating them before a tasting can bring your customer closer to your wine through surprise and education. This works particularly well if it is a couple and one of the two doesn't know wine at all, which happens often. Getting the significant other completely bought in only solidifies their total experience further.

If your license allows it, pair each sip with a very small piece of well-crafted food, maybe from a property-dedicated chef or local catering company. According to our demographic study, your customer is a foodie, and this pairing plays up to the notion that wine is a lifestyle event. They will love it and take home what they learn to share with others.

I'll keep this at the "unique idea" level and not go into the tasting experience details like pouring women first, showing the label, serving perfect glassware, having filtered water and a cloth napkin present, etc. I'll assume perfection is already in place if you've come this far.

5. ***To tour or not to tour? Here's a radical idea . . . let them decide***
    I believe at this level of a unique wine experience, offering a tour of the winery is more about showing off the property than teaching them about winemaking. The wine education happens at the sensory table tasting. If you have an amazing winery, give them the option to see it, but don't require it. Some of your customers live in homes nicer than the winery, so they may or may not be impressed. Or they have seen three wineries that day, and one more barrel will only send them over the top.

If they accept the tour, the winery must be 100% presentably perfect and the tour must be brief. If you're making the winery a part of the onstage experience, it needs to reflect the overall brand essence. I wouldn't want to have an awesome meal in a restaurant only to see a "working" kitchen that might change my perception of what I thought was a perfect place. Make it excellent . . . and make it brief. As cool as it is to you, most people get bored quickly.

Tour or no tour, you should invite your guests to stay as long as they like. The longer they stay, the more emotionally connected they become to the place and the people, thus the greater the chance that they will either feel obligated or want to buy wine.

Also, it's very important for your team to walk away at certain moments of the visit. People will feel freer to talk amongst themselves and influence the rest of the group without a wine expert hovering over the table. Your customer needs to privately express themselves to the party they came with to possibly justify spending more on wine. They can express how great the experience is or simply marvel at the surroundings. It will work to your advantage.

6. *The final impression and parting perfectly*
This experience should never come across as being desperate. Luxury by nature denotes confidence. If everything is perfect, the customer will buy. They may not buy today, but someday they will, and it's the journey you take them on that will impress them to buy in the end.

At parting, they should be presented a small order form on quality paper stock inviting them to buy wine. On it, denote which wines are available, with the number of cases next to the price if the remaining inventory is small. And denote which are either sold-out or allocated. For those allocated wines, give your team-member the power to "find" some obtainable bottles if they decide to buy that day. To get the wine continually, they are welcome to get on the waiting list or join the club.

While they are looking over the form, another opportunity presents itself to offer something unique . . . lunch or dinner arrangements. Make a deal with a local prestige restaurant that for a case or two of wine a month, you can call on them to provide your guests with last-minute reservations. This is a concierge-type service your team can offer, playing it up as something only your winery can do. The restaurant benefits by getting free wine to sell, as well as some exceptional clientele to add to their list of patrons. It's a win for everyone. And if the customer buys over $1,000 worth of wine at the winery, you can throw in a bottle for them to enjoy at dinner. Whatever you dream up, make it special and personalized.

Finally, many customers rarely enjoy traveling with wine, so kindly offer to ship it to them, if in a legal state, at a discounted rate or free depending on how much they purchase. When you send the wine, include a handwritten note from the team member who took care of them.

It's these types of personalized services that will set the bar very high.

7. **The Gift**
I had this happen at a 2-star Michelin restaurant, and it was one of the most impactful moments I've ever experienced. I was with my girlfriend at a special dinner, and at the end, the waiter placed a small gift box with a large, elegant bow on her plate. Inside was a tiny handmade truffle and a note that read, "Thank you for visiting us." It made an enormously positive impression and put a big smile on her face, which put a big smile on my face. Give this small gift to your guests on the way out. Chivalry is not dead, and even today in a world that has finally accepted strong women and gender equality, a woman should always be treated exceptionally. She will love it and he will love seeing her love it.

This was a detailed example for a luxury-type winery setting, so customize for your business model. If you are a rock-n-roll négociant brand that has a tasting room, offer live music, pair each course with a style of music, have edgier food and bolder glassware. But treat your customer like royalty either way, just in a rock-n-roll way.

If you instill some or all of these methods, without a doubt you will provide the finest wine experience available. Your customers will rave about you, buy from you, and tell others about you for years. It's all about the experience being unique and exceptional. And I hope these ideas will inspire you to create even more distinctive concepts to apply in your own way. Be a dreamer, be creative and shoot high. There was a day when an unknown dreamer named Walt Disney dared to do it as well.

The Coveted Wine Club

If a customer visits your tasting room, you have the opportunity to interact with and sell wine to them once. If they join your wine club, you interact and sell wine to them for years to come.

Offering a wine club is not a hidden secret. I would be confident in assuming that about 95% of the wineries offer some type of wine membership, and for good reasons because there are some tremendous benefits.

1. It's an opportunity to frequently stay connected with your most valued customers, reinforcing the brand awareness.
2. It is an additional and anticipated revenue stream, which is never a bad thing in the capital-intensive world of wine.
3. It provides a higher level of predictability and better forecasting for the sourcing and winemaking team when harvesting grapes, buying juice or creating wine.
4. It encourages wine innovation because the team may need to come up with additional wines to satisfy club shipment expectations.

Points #1 and #2 are really all you need to justify offering a club. It is one of the easiest decisions you will make.

And don't just think this is only for wine companies that have tasting room facilities. Offering clubs is particularly strong for virtual brands as well. There are some outstanding and quite large online wine memberships that have taken over the world recently. They all have their niche and offer everything from value to luxury wines, having a greater level of variety and member flexibility. As I mentioned previously, it is not a new initiative in the wine world, but a widely used way to brilliantly sell more wine and engage with your customers.

Keep a few important things in mind.

There are literally thousands of wine clubs available, so the competition for other wine brands to acquire your customer is significant. Next, the average member sticks with one club for roughly two years, then moves on to another. Finally, wine clubs are one of the first luxuries a customer will cancel in any economic downturn. When people have to tighten their belts, they immediately start with all the nice-to-have's in their lives, and wine club cancelations spike during these times. So, it is not without its challenges. However, the positives are just too *positive* to not fully engage in this excellent sales channel.

Once again, there could be an entire book written on this very important topic. However, in keeping with the spirit of this guide being brief yet concise, we will cover what you need to know to make the correct decisions when launching and managing a highly-effective club.

When developing and launching a wine club, you need to make some solid and important decisions upfront because once it is rolling, you really cannot change it later on. Your customers sign up with a known expectation of the ongoing cost, frequency and types of wines they will be receiving. Therefore, any changes in this expectation will only trigger a lot of cancellations at once. What I found is that most mature wine companies will have multiple versions of a club that was driven by the inability to change the original format, so they added variations of the club on as time progressed. Though this lessens some of the fallout of current members, it becomes an administrative nightmare for the team, raising costs and minimizing the team's efficiency toward other important aspects of the business.

When building a club, a few top priority decisions will need to be made. As with most anything in wine, this is a subjective list that can be up for debate because each club has its own perspective on what works and what does not. I found that being more conservative and staying "in the middle" will lessen chances of structural changes to the design of the club as well as keep most customers in the club longer.

The four most important decisions every wine business needs to make are flexibility, frequency, amount, and pricing.

- *Flexibility*
  Let's assume that you offer a fully customizable club where the members get to have a selection of whatever they want. If you had 100 members, you would absolutely have 100 different club iterations. Everyone has their specific needs and desires. But wine is not like most other industries because the product has extremely high costs, long lead times, and great complexities to make it. Sure, your memberships would be highly successful, but ultimately your much higher production and inventory costs would supersede any gains you get from having a greater number of members.

  For this reason alone, your club should be a fixed offering with very little to no flexibility. Be true to what you can do within your means. You may miss a few prospective customers, but the financial stability and predictability of your club will significantly outweigh those lost sign-ups. In my experience, customers will ask for the flexibility, but when it is explained that it is not a possibility, they will most likely sign up nonetheless.

- *Frequency*
  Look up any wine club online and you will see monthly, bi-monthly, bi-annually, quarterly, or "when the wine is ready, we will ship it" frequencies of clubs. I've managed clubs on

both sides of this equation. There is no magic formula to the answer, but I believe there is a middle path to consider.

Your customer loves your wine, so they want it around to enjoy. You, of course, want to sell them as much as possible. However, there is an inflection point where even the most loyal customers will cry uncle because they have too much of any one wine brand. It's the same as if you stayed at a party too long . . . they love you, but let's be honest, too much of you is too much. You want your wine to be accessible, but not too accessible. It must be special, and to have that perception, less frequent shipments are best. I have found that four times per year is ideal. It's enough so they don't forget about you, but not so frequent that they get tired of the brand or have too much wine at home. Also, you can have four solid shipments per year but offer additional *optional, members-only* deals when you have extra wine to sell. The members will feel very special that they have the option, so you'll pick up a few more sales, but those who do want the wine don't have to opt-in, thus staying in your club longer. It is a win-win, or wine-wine, right? Sorry, I tried.

Something else to consider is that shipping during the summer months can be risky because the wine has a much higher probability of getting over-heated in hot delivery trucks. This can also get expensive because they will need to be cold-pack shipments. Shipping four times per year will allow you to avoid these risks by scheduling the club shipments around the hottest months. Shipments can be made in February, early May, late September or early October when the heat breaks, and November. Each month has a benefit to its timing as well. For example, November wines will be perfect for Thanksgiving and Christmas seasons. February is when customers typically are past the gift-giving cost shock and realize they drank most of their wines during the holidays and need more as a result. May is excellent for white and rosé wines as the warmer days trigger this customer desire. And finally, October chill also triggers in your customers mind that warmth of great red wines as the winter approaches.

It's pretty perfect. Build a wine business that supports this type of club and let how you sell to be driven by the supply side, not your customers' expectations.

- ***Quantity***
Most wine companies want to sell as much wine as possible. For the majority, that means a lot of wine, but for some that can also mean a highly allocated, small amount of wine. These two scenarios are handled very differently.

  For the wine business that is scalable and has oceans of wine to sell, or when sales are healthy and you can get your hands-on oceans of wine, it's ideal to offer 4-bottle, 6-bottle and 12-bottle shipments. If your club ships more often than four times per year, lesser amounts per shipment would be better because the more wine a customer gets more frequently, the higher their risk of cancelation. Assuming it is the recommended four times per year, then 4, 6 and 12 are perfect. For a 4-bottle-per-week customer, which is fairly average for their demographic, it is well within their capability to make sure it's all gone by the next release.

  For the wine business that is boutique and has the perception of a higher level of luxury, less is more. Depending upon if your allocated and selling out, or if your wine lots are tiny, 2-bottle or 4-bottle offerings are ideal. It maintains the perception that the brand is highly sought-after, and also assists with managing the inventory so you don't sell out and miss a few members. But what if your brand luxury perception is very high and you're *not* selling out? The perfect way to handle this scenario is to offer a "wish list" to your members. A few weeks prior to the club shipment, send out an email to the members indicating the wine is coming and if they want more than their allocation, to send in a "wish request" that the wine company may or may not be able to fulfill. It is a great way to protect the brand identity of being boutique but sell more wine on hand.

- *Pricing*

    Look up any wine club online and you will see vastly differing pricing options and cost levels. The range is $40 to $1,500 per shipment. The majority tend to be in the $40 to $150 per shipment range, but the variations are endless, and this is mostly driven by the type of wine business and price points of the wine. This is a decision that is very specific to each business model for sure and there is no real playbook for what works and what does not. Too high and you will have a much higher cancelation rate. Too low and it will not feel as special, and you're possibly leaving dollars on the table.

    The first place to start is based upon an initiative we covered earlier stating you should know your COGS per bottle. Always know your costs before you price any wine for any reason. The margins in D2C are much higher than three-tier sales channels because they range from 50% to 75%. So if your bottle COGS is $20, then a possible price range is $30 to $35 per bottle. Again, this is a generalization as many factors play into this decision, such as your brands availability, prestige and luxury perception, scores and ratings, what pricing shows up in the wine apps and online if your wine is sold elsewhere, and so on. But the majority agree that 50% to 75% is a good range to stay within.

    Secondly, if you're using a membership-only business model, or if the brand is fully allocated and sold out upon release, or if your offering wines that cannot be found anywhere else but the tasting room, then no discounting is necessary. Don't go there, because you don't need to. The scarcity and high demand for the wine negates the reasoning to offer deals. And remember what we learned earlier about discounts - once you go there, you cannot come back. However, if the wine you are offering as a club shipment is also available to the general public either at the winery, online or at a retail store, then a members-only discount is imperative. What other reason would they have to join?

Then there's the dreaded shipping costs that all wineries and customers have to deal with. Compared to most products, the costs are high for two main reasons. We like to use glass for bottles, so the weight is heavy, and the package requires special handling by obtaining an adult signature from the person who receives it. Wineries mitigate this by offering discounted or free shipping rates, and I believe there is a way to handle this. Use discounted and free shipping as a carrot to get new sign-ups at the point of decision-making. Keep in mind that a 4-bottle offer has a much higher per-bottle shipping cost than a 12-bottle offer. The more wine, the lower the shipment cost per FedEx and UPS. So, little to no discounts on less wine, higher discounts or free shipping on 12-bottles and above. It's a balance between pleasing your customer and pleasing your finance team and bottom line.

The Even More Coveted Wait List

The even more coveted Wait List scenario is when your brand either sells out or you have an extremely high luxury perception. First, congratulations. Second, please don't screw it up because it is highly unique to have this situation.

Luxury wineries rarely refer to their allocated wine sales as a wine club. It is a wait list, kind of like waiting to get into your favorite restaurant because they are so popular and busy. It is really a brilliant marketing tool because it appeals greatly to your luxury consumer. I believe all wineries selling a wine over $100 per bottle need to have a wait list, even if there is no wait. You might have to reread that sentence to get the full meaning behind it. The perception of a customer waiting, and then being accepted, is more powerful than automatically being accepted. It creates an air of exclusivity and scarcity, which should entice your luxury customer. However, a wait list that seems to go on forever is as counterintuitive as not having one at all. I strongly believe your customer needs to have access at some point, or it can work against you. They will move on to other wine brands if this is not handled properly.

Take a bold approach. Each time someone asks to be part of the allocation process upon new releases, have your team mention there is a wait list and they would be happy to add them. But because they are in the winery they can purchase anything they'd like to take home. That's the bait and the wait. Next, if there are spots available, meaning you have more wine than the demand at that moment, stay in touch with them via email over a series of weeks. Send out a series of emails with an update of where they are on the list. For example, if there are 20 spots available in August, you can let them know with each email their progress from 19, to 12, to 4, etc. Finally, announce to them when they are "in" and process their allocation. With this method, again, you create the perception of exclusivity and scarcity. It's extremely important to offer them the wine at some point so there is final customer gratification.

If there are no spots available, meaning your demand actually outpaces your supply, then I suggest something slightly unorthodox as well. If you have 1,000 people on your waiting list, wouldn't it be ideal to know who the most important and coveted people are, those customers that purchase the most wine and have greatest loyalty? Having the ability to "tier" the waiting list is a highly desirable proposition. Upon the customer asking for an allocation, your team inconspicuously asks them a series of question about their interests and activities, how often they purchase wine, how many other allocations they are a part of, how often they drink wine in your price category, etc. With this information, internally, you can give them a "grade" that will either put them ahead of most if they fit the description of a great customer, or further down the list if they don't. No one has ever said that a wait list has to be first-in, first-out. It can be all-in, most valued and committed customer out.

The perception of a wait list is at times more powerful and impactful than actually having a wait list itself. Handle it carefully and calculatedly because it is the D2C promised land once developed.

The Occasional Yet Powerful Use of Email

Think about how many emails you receive in the morning. I'm fairly sure there isn't a company that emails you frequently and you love it so much that you ask for more. Our inboxes are inundated with offers, deals, scams, questionnaires and spam. Then there are the real emails from your work, family and friends. It is just too much, and you don't want your company's emails to add to this misery and be dragged to the junk pile by potential customers.

When considering the use of emails, there are two very different types of wine businesses. Basically, those that have wine generally available to their customers, like out in the marketplace, and those that do not, like the wait list scenario outlined earlier.

For those with wine that is more accessible, your email list becomes mostly a source of information. Your social media platforms should be used for ongoing updates, interesting happenings, and generally just keeping in touch to stay at the forefront of their minds. Email should be used when there is an important reason to connect or a call to action that is only being offered to a select few of your customers, or at least that should be the perception. Email touch points should be infrequent and always personalized, if possible. They should be used for announcements such as a special wine release, new partnership offering, or an exclusive invitation, such as a winemaker visit to their area. It is more than okay to put in a special wine offer or an invitation to join the club on occasion, but if the messages are too heavy handed or too frequent, the emails will have a substantially higher unsubscribe rate. Your customer will look forward to reading your emails if they are educational, meaningful and occasionally have something to offer. Anything different and your emails are at risk of hitting the junk pile.

A creative idea is to use email to offer something very special and unique to your customers. For example, your winery can offer a once-in-a-lifetime exclusive wine dinner at your customer's home for a number of their friends, catered by a top local chef and attended by your winemaker. Again, appealing to your specific customer demographics, controlling the sales environment, and being associated with other luxury products exposes you to a larger key group of customers. Make the fee high because you only want a few to take this offer. Remember, the level of excellence you will be offering will not be inexpensive. And keep it super-exclusive. Most importantly, if no one bites, the email comes across as a perceived luxury-level offer intended for only a few select customers. Most of your customers who don't take the offer will see that your brand can offer exclusivity of personal experience that is above the rest, and the rising tide lifts all boats once again, just for the cost of an email. This is a luxury winery example, so take this idea and make it your own as it pertains to your unique brand essence.

For those with wine that is not generally accessible to their customers and no club is available, then email becomes much more of a selling tool. Firstly, I don't believe there should be an online store for general purposes. It appears unallocated and proves "anyone can get the wine." In terms of email usage, I still strongly believe in keeping the touch points very infrequent given the high volume everyone experiences in a day. However, they can be much more about selling wine. Your customer is conditioned to look for these opportunities because they cannot get your wines any other way. Upon each release, send out a perfectly-crafted email that is very much like a billboard in terms of content. Get to the point quickly and concisely. And make buying easy, like one click away. Even if they are highly loyal to your wine brand, they still want an Amazon-like shopping experience or they will quickly bounce out of the sale. To base a large portion of your sales revenue on an email-only type of business model is very precarious because there is no predictability from one release to another. So it is always best to offer a club-type membership to your followers, and use email as an infrequent yet powerful marketing and sales tool.

For all of the reasons outlined earlier, and many more you will find out in due time, I'm sure you agree that D2C is your field of diamonds waiting to be discovered. It is brilliant for so many reasons, particularly the direct customer touch point, the higher margins, and the opportunity to communicate your awesome brand messaging. The higher the percentage of your D2C sales, the higher the probability you have for success, and money. It is a one-to-one relationship, so take it on with excitement and excellence!

So, You Want to Play the 3-Tier Game?

Before diving into this section, this is where the path will diverge, and it is time for decision-making once again. If the goal is to keep your wine company boutique or niche, then D2C should be your primary marketplace with a very focused strategy. The benefits are just too significant. If you have selected this business model and you're still considering using the 3-tier network, it is always best to be extremely cautious as you make the final decision. The vacuum of the 3-tier will take you, your valuable resources, and your dollars way further than you may anticipate. This is not a world where you can be half-in and succeed. It is inevitable that whomever goes here will be pulled in further to the all-encompassing responsibilities once they get rolling with a few sales wins. So, it may be best to focus and build an outstanding D2C-based business. Then, once the company matures, has a higher level of cash flow, and acquires more expertise and better understanding of how the brand is being embraced by the customer, it may be a good time to consider dipping your toe into the murky 3-tier waters on a very limited basis. If this was not enough to persuade you and you still want to see your wine in your favorite local restaurant, then the best way is for you to engage in the broker network. Like a stockbroker, they are the liaison between you and the 3-tier network. The great ones do the work from top to bottom and get a commission of anywhere between 10% to 20%. Considering the fact that you are already working off of small margins in the 3-tier world, it is an expensive proposition, but effective if you really feel the need to be distributed as a boutique wine company.

If your model is to scale your business to larger volumes, then this is a must-read for sure. No one likes to jump into a warm pool only to find out it is actually freezing. You should be very aware of this plunge, how to play the game, and ultimately, how to set yourself up for the greatest level of success.

To begin, this section is aptly titled because the 3-tier system in the US is very much a game. I have spent many years and over a million flying miles visiting just about every state, every wholesaler, and every retailer known to mankind. I've been on sales calls when it was 20 degrees below zero in Cincinnati and when it was 85 degrees on a beach in Waikiki, HI. I've been in an underground military bunker that was recently turned into an office space, and on a private plane that escorted me to eight different cities in a matter of a few days. I have the stories, the successes, and even more scars to know that the 3-tier system is not fun and is the very definition of "the grind". But it also can make a wine brand a tremendous success.

Just in case you are a super newbie to the wine business and not exactly sure what this is, I'll briefly explain. This is when you sell to distributors or brokers around the country that then resell to on-premise (restaurant) or off-premise (wine shop) retail accounts. It is the most important difference if your business strategy is to scale higher, as this is really the only way to gain those higher volumes necessary to achieve your goals. Simply put, the 3-tier system is a network of wholesalers and retailers who will sell your wines out in the marketplace. Some companies are specific to particular states and comply with the laws in those states, and some do not. Some are huge, multibillion-dollar operations, and some are not. Some use brokers and importers to obtain wine, and some do not. Some sell other people's brands, and some do not. Some play by the rules on a fair playing field for all wine companies, big and small, but most do not. Simple, right? I'm being lighthearted here, but those statements are largely true. It is an old, highly-regulated system that has both the federal and state governments imparting their influence, opinions, laws, religious beliefs, and party affiliations down on our industry for decades. If your simple wish is to sell wine to Whole Foods in Illinois or your stepbrother's pizza place in New York, it is an extremely expensive, time-consuming, sole-crunching and difficult part of our business. Yay.

Why do wine companies use the 3-tier system then? Because the exposure and the possibility for high, scalable volumes is huge. It's not because of the margins, which can be as low at 20% compared to the 50% to 70% of D2C. But the potential for greater volumes exposes the brand to many more customers, vastly improves the costs of buying wine and packaging, and greatly reduces the costs in the cellar as a full winery is a much more efficient winery. It is a very enticing market segment to sell wine for sure, but you cannot take it lightly or treat it as a side business. If you go to war, you need to have the capital, time, patience, perseverance and talent, which is not inexpensive, to achieve your goals.

The wholesalers that make up the 3-tier system have two extremely important characteristics you should know well.

1. *They make most of their money on the larger brands*
   The majority of their revenue is generated off the larger, more established brands. The retailers rely upon them for higher sales, so naturally the wholesalers focus their resources and team on these brands much more. They do believe that having a diversified portfolio of mid-sized and smaller brands is important, but these wholesalers are driven by daily sales goals which normally can only be achieved by the larger wine brands. It creates a very challenging marketplace for new brands to enter and gain a foothold.

2. *Your brand is just not that special to them, but there is a good reason why*
   They have an enormous amount of wine brands, sometimes thousands, that they have to sell to survive. To expect them to get excited and have their enthusiastic focus on your cool brand concept, even if it is over 50,000 cases, is just not a reality. If you meet with them, you'll most likely get a warm-and-fuzzy, but the other twenty wine companies that visit that week will get the same treatment. It's about keeping us wine companies happy and then conducting their business in the way that will generate sales, which is usually the larger brands. On occasion, you will get a wholesaler that is willing to help move the needle of a smaller to mid-sized brand, but I have found that is the exception rather than the rule. If you are lucky to find that partner, hold them close and don't let them go, because they are special.

And we're all big boys and girls here, so this reality is fine as long as you don't set your expectations too high. Know what it is going in, treat them with respect, know you understand their position, and get ready to play the game, because it can return great success if done well.

Having the right wholesale partners can make all of the difference and can either be debilitating or liberating to your plans. Though some are more capable than others, assuming their sales team will start selling your product like wildfire is probably one of the most misunderstood and disappointing aspects to new wine ventures. Until they prove differently, you will find wholesale partners exist to move cases from one place to another legally, deal with daily customer requests, and collect payments. They will very rarely, or just not at all, sell your brand proactively and regularly. You can give them an incentive for a period of time that will sell a few extra cases, but I assure you they will move on to other brands soon thereafter.

So, who does the actually selling of your brands and closes your most coveted and sought-after accounts?

If you partner with a company from the wholesale or broker network, logic dictates it should be them. Yet most of the time this is not the case. The onus falls on your company, especially as the accounts get larger when referring to the scalable strategy. So, you will need to have a sales team that will have to do most of the selling, which requires talented people, including you, willing to travel and sell from place to place. The more your team is out in the market, the more you will sell and the healthier your business will be. It's a fact.

The other common misunderstanding is the level of your continued financial investment beyond the one in your own brand and team. All wholesalers and most retailers expect some level of "marketing support" dollars that come in the form of incentive programs, discounts and marketing materials. The incentives could be to motivate the sales team, which is ironic because you're paying to motivate a team that will ultimately make commissions from selling your wine. Or the incentives could give the representatives "tools" to entice retail buyers to bring in your wine. An example of this is shelf discounts or quantity purchase discounts. So, you will need to establish a healthy budget for most retail accounts that your wine sells within. Typically, this could cost from $12 to $36 per case, which really begins to eat into your margins. Let the games begin.

There is an easy way to understand your additional investment as it relates to your long-term strategy. The more cases you want to sell, the more investment it will require, especially if you're dealing with larger retailers. The less cases, the smaller the investment. From a financial perspective, it's that simple. The goal is to sell so many cases that it will supersede the investment you put into it. But I warn you, sometimes it takes years to get to this inflection point.

Great wine is not enough to thrive and succeed in the 3-tier network.

6 Imperatives for Establishing a Successful Distribution Network

They call it the 3-tier network because there are three levels of companies and people that require your full attention and strategy for each.

1. The wholesaler
2. The retailer
3. The customer

We spent a fair amount of time covering your relationship with the direct customer in the previous marketing sections, so here we will focus on the wholesaler and retailer relationships and management. Not only do they make up your distribution network, which has a considerably high cost associated with it, but they also are required by law in most states. There are some states where, with the correct licensing, a wine company can sell directly to the retailer, as in California, but this is the exception rather than the rule. The majority of states have a very solid 3-tier network that have massive lobbying powers, so it is not going away anytime soon. And the games play on.

Before defining the steps of how to establish a wholesale network, there are a few types of wholesale partners to be aware of and consider. Who you choose to partner with ultimately depends upon your long-term strategy.

- ***Large National Wholesaler***
  The alcohol wholesale industry is extremely influential and powerful. Some would say the most powerful companies in the business truly run the show in most states. There has been considerable consolidation recently, which has shrunk the truly national players down to just one, that being Southern Glazers Wine and Spirits. They have a massive book of products, suppliers and customers. They require a considerable amount of resources and time to work with, but they can have the largest impact on your business.

- ***Mid-size Regional Wholesaler***
  This is the most significant portion of the wholesalers; mid-sized companies that cover a few states typically adjacent to each other. They have significant reach in the states they cover, large books of suppliers and retail customers, and excellent relationships with the top buyers given their regional expertise and focus.

- ***State & County Wholesaler***
  Filling out the country is a considerable number of smaller state or even county-level wholesalers. They are fairly small companies that sell to less populated states where the competition from the larger wholesalers is minimal. Or, if they are in larger states, they either cover counties well out on the fringe or have a very focused niche purpose; for example, high-end CA wines, imports from Italy, private label wines for on-premise accounts, and so on.

Being aware of and understanding the wholesale network is important. Partnering with someone who is a perfect match is imperative.

If you are new to this and establishing a distribution network for the first time, the following steps are a guide to get you started in the right direction. Also, they work extremely well if you're launching a new brand or looking to rework an existing network already in place.

This is a good time to go refill that goblet of wine once again, then continue. We will wait for you.

1. ***Have a firm sales and marketing plan in place***
   Before going to one wholesaler meeting, you'll need to have your sales goals firmly established and your marketing materials ready. Wholesalers like to work from facts and a firm direction. Also, this creates an understanding that can be easily measured by your team long after you leave.

   Keep everything very brief and simplified. These companies have an enormous number of tasks they need to fit into one day and will not remember details beyond the top level. Try to get everything on one page. Yes, just one page. It is possible. Some million-case brands have been able to communicate their sales and marketing goals on one page, so you can definitely do it as well. And this will save you time because most pages beyond that find their way into the round filing cabinet on the ground once you leave a meeting.

   Some ideas on goals and stipulations would be the number of cases you will be allocated for each release, or the exact profile and even the names of the retail accounts you would like to target. We will go into detail on how to accomplish this later. Also, clarify the pricing structure and range you are allowing them to negotiate for each deal and give them a clear marketing budget for incentives or programs.

2. ***Keep your target markets focused on a core group***
   A mistake that many new wine ventures make is to spread out and enter into whatever markets that will have them, which usually includes CA. This is a serious misstep because the more your resources are spread out, the less attention and love you can offer each. Each state you select to distribute in will require a significant

effort to manage and visit often, so it is always an excellent idea to choose just a few and really focus, showing them the commitment and love they require. As you will learn later on, the majority of the responsibilities are on you and your team, so focusing is key here.

Pick your top target markets, which are usually cities in CA, NY, IL, FL, TX because these five states cover the majority of luxury wine sales. As your business matures and broadens, it's okay to focus on secondary markets like major cities in NV, AZ, NC, etc. But to start, stay focused and put your limited resources into the top five to eight markets. This will benefit you in many ways.

3. *Find and meet with prospective partners*
It is fairly common to date before you get married. The business relationship you will enter into with these companies can be long and detailed, so you want to make sure as much as possible that you are partnering with someone that is passionate about your brand, capable of reaching the goals, communicates extremely well, and has a good cost structure so your wines can be competitive in the marketplace.

To have the highest probability to partner with this perfect match, ideally you will meet with at least 3-4 potential candidates in each market. There are two ways to find them: a) call a few restaurants in the marketplace that are a perfect future customer for your wine and ask them what wholesaler partners they work with, and b) look at some of your competitors' websites because they may post who they are partnering with in each market. Using both of these methods, you should be able to find a few of the strong players to meet.

Next, schedule a visit to the market and conduct interview-type meetings. Ask questions about their ways of working, company goals, approach, team expertise and clientele. You'll get an overall sense of their commitment to managing your brand at a very high level. You should be looking for a well-established, sophisticated partner in terms of wine knowledge, and has customers that match your brand positioning goals. They need to gain your trust and exceed your expectations in terms of your pursuit of excellence. You also need to have the overall feeling that they are a cultural fit that can withstand the ups and downs, with no BS when times get tough. This is a business marriage, so be picky and don't compromise.

With that said, don't set your expectations too high because only you and your team can represent your brand at the highest level. You're looking for a partner who can come close to the top, be committed, be passionate, communicate well, and take great direction.

4. *Shake hands and establish a one-year courting period*
   After you've done your due diligence, it's time to partner with the selected few lucky enough to represent your special brand. Sit down with your team, have a collective discussion considering all the options, facts and intangibles, then make your decisions. They will come down to mostly intuitive choices. With this level of due diligence, which most wine companies do not do, your chances of overall success will be high.

   If possible, you should not sign any long-term agreements at this point. Though things may look good at first, most wholesaler partners' true colors will emerge shortly thereafter, usually within a year. You will find out quickly that some lock in and engage quickly, and others were more promise than action. Have a one-year Letter of Intent, a soft agreement including goals, roles and expectations. At the end of one year, you will definitely know if they are the right partners for your wine company.

5. *The 95/5 rule: Find the few representatives that do most of the selling*
   Here is an excellent way to find your top hunters who can make the most difference to your success in each market. This will happen, so be on the lookout right from the beginning of the partnership.

   At some point, your team should get a call to present new wines and the brand story to the wholesaler representatives at one of their sales meetings. As they present, the majority of the wholesaler representatives will look incredibly disinterested, like they've heard and seen it a million times before. Someone's head could explode, only to have the representatives be annoyed by the mess and quickly move on to the next wine company presenting. I wish I were kidding, but I'm not. Take it from someone who has presented hundreds of times - this is consistent across the country. It's humbling to travel so far and sell your heart out in front of them only to get the feeling that no one is listening. It won't make sense to you because they benefit as much, if not more, than you do from selling your wine. However, remember what is in their minds and pressuring them most days. Big brands make sales goals fast and smaller brands, especially new ones, are a larger effort that may not pay off.

Here is a great way to play the game and find those diamonds in a room filled with rocks. I have discovered that in every presentation there are always a few wholesale representatives in the audience that take interest, ask good questions, and stay engaged. I've seen this in a room of 90 or even 11 representatives, and it holds true every time. Certain people are drawn to your brand presentation for whatever reason. You can see them while you're presenting. They will be engaged and possibly asking questions. These are your rock stars and require all of your team's time and energy. When this happens, identify who those representatives are and make them your best ongoing friends. Out of a team of 20, having one or two who proactively sell your wine is what typically happens, so find those representatives and get very close to them. Assign your top sales team representatives to them and have them stay in touch no less than once per week. Give your team the freedom to support them however possible. Consider them an extension of your sales team. Spend very little time or effort on the rest of the wholesale team in that market because it will largely be a waste of time. Focus 99% of your time and energy on those who support your cause. You will find that these 5% of the wholesale representatives will do 95% of the selling you will need to hit the goals.

6. ***Have firm short-term goals and a target account list per market***
One part of this, all wine companies do. The other, almost no one does, and I think it is a huge way to win in the wholesale network game.

Depending upon your overall strategy, have case or revenue targets that are no more than six months out. A shorter timeframe is not realistic because it takes time to get new retail listings closed and the wine shipped, and further out is fairly unpredictable because the six months leading up to that period of time heavily influence the trends and numbers. Having a six-month plan can be very realistic and measurable, and something that all sales representatives will most likely get behind. Have your team use these per-state goals frequently and consistently with wholesale partners to reinforce the fact that you are on top of your business and have firm expectations.

This next initiative is the game-changer. Have your top wholesale representatives provide a target list of accounts in their state that will be the core prospects for your wines. Most wine companies, if not nearly all, hand over volume or revenue goals with the expectations they will be executed perfectly, which almost never happens. To make sure there is a high-level of excellence in the marketplace, and that the representatives are hunting in the retail accounts your brand is perfectly positioned for, an account list is the only way to go. They may push back, but that means you need to push forward harder to get it completed and managed regularly, because this list is your golden egg. At the end of a mutually agreed upon period of time, like three months, have your team go over it account-by-account to determine if the plan is unfolding well or if there are issues. It is putting it all on the table. Without it, wholesalers can hide the details behind blanket statements if they do not achieve their goals. And, most importantly, powered with this knowledge, your team can visit each market and go on sales calls to help the wholesaler reach their goals. It is an imperative exercise and a total win-win when done regularly and done well.

4 Engagement Imperatives of the Wholesale Network

Wholesaler representatives have a tremendous amount of relentless activity and work within a very high-pressure sales environment that is primarily number-driven. They have to survive in exceptionally competitive environments because a "normal day" for them is to achieve daily sales goals, fight the competition, vie for the attention of retail buyers, work with needy suppliers (you), and balance the internal pressures from their executive team to hit their goals. They are extremely competitive, type-A personalities who are born to win at almost all costs. As you can imagine, this inherently forces them to give the most attention to the brands that will achieve these goals. This doesn't mean your brand is not important to them, it just means your brand isn't important most days, until the day it has to matter to achieve sales goals.

So, how does your team manage a very busy, rather indifferent set of wholesale partners? It is fairly simple if you follow these rules of engagement and management.

- ***Be perfect and buttoned-up all the time***
  Given the nature of their business, any wine company that comes to the partnership being perfect all the time makes the wholesalers life much easier and better, so they will appreciate this tremendously. Going into meetings or conference calls, have your team make extra efforts to gather the information required, be extremely well-versed in your business in terms of pricing, wine and operations, and have the volume goals and accounts lists pre-created and perfectly updated. This requires your team to spend quality time preparing and practicing well before they engage with them. For important planning and strategy meetings, sit down with your team well beforehand and discuss the plan of attack, materials needed, whether or not other team members need to contribute data, and so forth. Emphasis with your team that it is not an option to go into wholesale meetings winging it. It reflects poorly on your company overall and signals to them that they should not meet with you frequently, and to spend much less time on your brands. The wine companies that are the most buttoned-

up will get more attention and effort from an already distracted wholesale network.

- ***Be brief***
  From every one of the tens of thousands of representatives I've come across and the thousands of meetings I've held, there was one major common element. They love brief and concise phone calls and meetings. Even with your extensive planning and initiatives, they all need to be distilled down to one, or on occasion, a few pages. Everything from sales goals, marketing information, account lists, etc. You hand them one page they can have with them and refer to at all times, and it's like handed them gold.

  Try this other tactic just one time and you'll see from their reaction how perfect it is. When your team gets on a call or goes into a meeting with a wholesaler, immediately have them set the timeframe. Make sure that their engagement will be brief and then hold very firm to that guarantee. For example, walking into a meeting, the team should start by announcing to everyone it will be a 30-minute or 10-minute call, depending upon the type of situation. Immediately you will gain the respect of the wholesale representatives for showing respect and understanding of their worlds. This is very unique because no one else I've ever seen has does this well.

- ***Be diligent about check-ins, meetings and responses to last-minute requests***
  Most of your wholesaler partners will be located throughout the country, so they will not see your team all too often, which is fine because they don't want to see your team all too often. However, "out of sight thus out of mind" does happen and the wholesaler will get focused on the shiniest objects in their universe at any given period of time. Therefore, contact points are imperative to keep your brands at the top of their minds, review goals, check progress, and generally let them know you're thinking about them and there to help.

To not be a burden or completely forgotten, I suggest the following. Have your team call your top wholesale representatives at least once per month. This will include, at minimum, every representative that made up the 5% outlined earlier. Next, select your top 8 to 10 markets and plan on visiting them at least once every six months to conduct more in-depth strategy meetings where you can go over the volume and account list goals, discuss issues, come to an agreement on the next six months' goals, and break bread to say thank you for being a big part of your extended team. These are the 30 to 45 minute meetings where you need to be perfectly buttoned-up. Come in, be great, get out. They will respect you for it.

Also, when a wholesale representative calls or emails your team with a request, they should all be treated with the utmost urgency and answered within hours, not the next day. Even if your team doesn't have the immediate answer, they will appreciate knowing someone is on it and will be responding back shortly. If they reach out, that usually means there is an opportunity ready to go down and they need some final answers to move it forward. However, be careful not to chase every opportunity they approach you with. Stay within your brand essence, price points and goals. It is very enticing to try and get all the deals out there; however, saying no to the ones that are not ideal for your brand is difficult yet important to the overall health of your financials, as well as the relationships with your representatives out in the marketplace. Do what you know can be executed upon perfectly and pass on the rest. It will reward you in the end.

- ***Be a sales-driven culture and sell to the top prospected accounts***
  Finally, when determining your own faith, nothing works better than this recommendation.

Have your team go into the marketplace unaccompanied to target and land the top opportunities on the account list. You might think the wholesale representative would feel this is encroaching upon their turf, but as long as you notify them of all the details, they will very much appreciate the effort and most likely assist in some ways. Someone else working their accounts to try to land deals that they will benefit from is seen as the ultimate relationship builder and win-win.

I have learned from multiple different situations that having a sales team that only manages a wholesale network, and then tasking their representatives with selling and making your goals, will fail more times than not. This is not taking your success into your own hands, it's just relying upon others and hoping for success. Jess Jackson, the legendary owner of Kendall-Jackson, preached this very often. He knew early on that the only way for his chardonnay to be everywhere, and at the price points he was expecting to protect the brand, he would need an internal army of excellent sales representatives who would make direct calls to accounts, land them as a new customer, and hand them over to the wholesale partners to supply and manage. For a wine that has experienced nearly 90% distribution to all retailers that offer chardonnay between $10 and $15 dollars, I am going to be a follower of this methodology for sure. Being the Brand Manager of this renowned brand, I saw how it worked firsthand.

The process of prospecting and closing accounts is an artform your team will need to not only embrace but be excellent at if you are willing to play the 3-tier game. It is an expensive and time-consuming proposition, but without it, the chance of success is very minimal. It is the cost of doing business at the highest level in the wine industry.

The final aspect to understand about your network of partners may not be terribly encouraging. It is not a two-way street.

Given the complexity and outdated nature of alcohol laws in the US that were mostly established at the end of prohibition in 1933, the 3-tier system is here to stay for the foreseeable future. We all nitpick at it, often over pints of Lagunitas A Little Sumpin' Sumpin' Ale, but at the end of the day it is a network of partners that should be understood and approached with cleverness about how to handle it most effectively. With the boutique strategy, they are driven more by passion and love of wine than volume, so expect an easier, more personal working relationship. With the scalable strategy, it's a bit tougher given the massive levels of competition as well as marketing support dollars flowing about from other larger wine companies. Either way, make sure to hire a sales team that has experience in the types of wholesalers you will be partnering with. Don't set your expectations or your growth plans too high. Be realistic, knowing your team will have to do most of the heavy lifting with selling accounts and staying in touch. Even with this wholesale knowledge, you will find it challenging and frustrating, but it's necessary, so get thick skin and learn to power through. Come to terms in your own mind with this notion and be okay with it.

Welcome to the wine industry and the glorious 3-tier distribution system.

Defining Your Target Marketplaces

When we think of selling wine, we typically think of the more interesting retail stores that make large, exciting purchases such as Whole Foods, or the romance of fine wine sales to the 3-star restaurant The French Laundry, but selling wine happens every day in places you wouldn't always imagine. What about smaller but prestige regional grocery chains found in all parts of the country? It happens, often. Or selling wine D2C online. This happens with even more frequency as of late. The point being, though it's all "just selling wine", it does take different levels of knowledge, experience, personalities and talents across these varied and many different sales segments to be successful.

For scalable wine companies, each segment typically has individual sales representatives who have experience and talent in the respective segments they are selling into. For boutique wine companies, a representative has multiple segments, or sometimes all, and thus needs to have a broader skillset and general experience levels.

Once you have a good understanding of the different retail segments, we'll move onto defining the specific types of wine sales professionals who are ideal for each.

Off Premise vs. On Premise:
These are well-known industry terms used to describe any retail licensed establishment that purchases wine from a winery, wholesaler, negotiate or importer and sells to their customers who either take the wine "off the premises" to drink (Off Premise) or stay "on the premises" to drink (On Premise).

- *National Chain Accounts*
  Off Premise Description: A retailer, usually large grocery stores, that has broad, national reach with many locations, sizable wine sections, significant buying power and high wine sales. They can have multiple "banners" or store names. They represent the mass majority of wine sales in the US overall. The term "stack em' high, stack 'em deep" is not uncommon when talking about their business practices. Off Premise examples are Costco, Sam's Club, Whole Foods, Trader Joe's, Kroger, Albertsons, Target, Walmart, 7-11.

  On Premise Description: A restaurant chain, hotels, airlines or cruise ships that have broad, national reach with many locations, significant buying power and wine sales that can either be very high or alarmingly low. Frequently, they have limited menu options and usually lower-end wine programs. Think about the last time you visited Applebee's or Olive Garden. Small wine list with a few options, right? They can have one or multiple banners under one corporate flag with central buying power. Examples are CPK, Darden (Olive Garden), Cheesecake Factory, Applebee's, Hyatt, Marriott, Carnival Cruise Lines, Royal Caribbean Cruise Lines, Delta, American, United.

- *Regional Chain Accounts*
  Off Premise Description: A retailer, mostly grocery chains, that have broad regional reach, usually within one state or across a few adjacent states, large wine sections, medium to large buying power and significant wine sales. Usually under one banner name. Examples are HEB, Publix, Foodland, Raley's, Giant Eagle, Sprouts, Winn Dixie, Stop & Shop, Piggly Wiggly, Stater Brothers, Fresh Market.

On Premise Description: There are very few, if any, regional restaurant chains. They fall into large national accounts, smaller specialty accounts, or broad market categories outlined in other sections here.

- **Specialty Accounts**
  Off Premise Description: A retailer that has a particular specialty, like liquor or wine-only sales, one state or across a few adjacent states, large wine sections and medium buying power. Examples are BevMo, ABC Fine Wines & Spirits, Specs, Binny's, Bottle King, Liquor Barn, Brown Jug, ABC Stores Hawaii, Vino Volo.

  On Premise Description: A chain that has broad regional reach, usually within one or across many states, that have a very focused specialty in one area like a steak house or Italian cuisine. These can be either lower-end themes with short wine lists, or more often, high-end fine dining themes with longer and more sophisticated wine lists. Examples are Maestros, Il Fornaio, Longhorn Steakhouse, Maggiano's Little Italy, Bahama Breeze, Ruth's Chris, Morton's, Bubba Gump Shrimp Co, Eddie V's, Seasons 52.

- **Broad Market Accounts**
  Off Premise Description: Basically, every mom-n-pop liquor store, gas station or local grocery store you can imagine. Big, small, nice or funky, they all qualify. It's the catch-all bucket for everything that is remaining in retail. Examples are . . . put down the wine glass for a moment, walk out of your house or office, and look down the street. You should see a few.

  On Premise Description: Again, basically every mom-n-pop restaurant, or small cheesy and luxury hotels from Hawaii to Florida. Examples are . . . are you hungry tonight? Chances are you'll visit one then.

- **Luxury Accounts**
  These accounts technically can qualify for Broad Market but are very specialized and thus require a separate category. They ordinarily are one to two locations and focus on either small wine

producers, higher-end or extremely high-end luxury wines, appealing to a more sophisticated buyer looking for special wines from a knowledgeable staff. For this reason, I separated them out given their unique and high expertise needed by the sales representative selling wine to them.

Off Premise Description: Luxury wine shop with wines ranging from $15 (yes, that is luxury in most parts of the US) to thousands of dollars per bottle. Caters to a high-end clientele with special services and unique wine offerings. Some examples are Wally's Beverley Hills, K&L Wine Merchants of CA, Acker Merrall & Condit New York, Sherry-Lehmann of New York.

On Premise Description: Luxury restaurants, lovingly referred to as White Tablecloth, with diverse and interesting BTG (by-the-glass) lists ranging from $10 to $50 per glass, and bottles from $50 – thousands of dollars, hopefully with an outstanding chef-driven menu. Reserved for 3 and 2-star Michelin rated restaurants, or similar experiences (country clubs, luxury yacht services, private jet services). Examples are The French Laundry of Napa, Per Se of New York, Alinea of Chicago, Gary Danko of San Francisco.

- ***Direct-To-Consumer / Online Accounts***
  With the state laws continuing their post-prohibition, old school ways and approving wine shipments direct to your customers' doorsteps, this is a fairly new category but is establishing a significant amount of sales, in the billions of dollars when this was written. They "feel" like a large, national, off premise account but have very unique needs depending upon the nature of their business. Exciting and a new shiny penny to play with. Examples are Wine.com, Vivino and many others.

A considerable amount to digest, I understand, but knowledge is power and the more you know, the better decisions you can make when building an outstanding sales team. Based upon your marketing positioning statements you created, it is imperative that you clearly define what retail marketplaces you will focus on and target. "All of them" is not a realistic list because the diversity of selling techniques, and the difference in sales team skillsets, are just too vast. Focus on a few segments and success will come much more readily.

Building a World-Class Sales Team

As we have learned, your team will do most of the heavy lifting and in-person sales when it comes to selling to the most important retail customers, regardless of your strategy. No one can sell your brand and story as well as you and your team. Your customers, if they are direct or via retail, are your customers and need to be heard and supported by you. I've managed many types of sales teams and learned numerous lessons. Sales is the single most important role to the very existence of all the expectations, the current company goals, and any goals that come up in the future. With great salespeople, a wine business can continue on the journey toward whatever it intends to achieve. This could be as cool as getting more people to just taste your wine, or simply increasing cash flow so you can improve winery equipment, open new markets, hire more talented employees, pay bills, pay loans, pay investors, and fuel an already successful business to new heights.

It's sales that right all wrongs. This will require you to have a world-class sales team. So, to understand this very complex and essential role much further, let's start by defining the characteristics and personality types you should be considering.

- *The fear-driven sales representative*
  Finding a source of motivation is important for most sales representatives. Wine sales is a daily grind that requires a kind of reason-to-believe that drives them to get up every morning, celebrate or brush off the past, and move the ball forward like it's the first day on the job.

  Regardless of talent, experience, capabilities and intuitiveness, there is a type of sales representative that is motivated by constant fear and it drives their daily activities, usually in the wrong directions. They have constant fear of losing their job, fear of being looked down upon, fear of not achieving, and fear of disappointing others. Whatever way that fear manifests itself, it quickly becomes apparent when working with a fear-based sales representative. The decisions they make, the answers they provide, their disposition in meetings or electronic communication, they are all slightly skewed by this one element. And if the internal team can see and feel it, it's always true the customer can as well. Fear-driven activities create frustrations and biases, and do not create long-term relationships. Fear of failing is not an effective motivation, and it's clear to others, which only sets these sales representatives back from what they are trying to achieve.

- ***The confidence-driven sales representative***
  If confidence is their motivation, that is an excellent start, but it is important to be careful here as well. There is a very thin line between confidence and arrogance. George Clooney is confident. Yes, at times he can push that line, but he comes across as a smart, caring, knowledgeable person that you feel you know already without even meeting him. Representatives who have this trait just feel like someone you can have a glass of wine with and talk for hours. This is the X-factor.

  They are motivated by their need to achieve, succeed and provide value to others, either customers or internal peers. Though they have the same fears, worries, dreams and financial responsibilities as everyone else, they react to their daily grind with a higher reasoning, a purpose beyond just fear of failing. They care more about their reaching
  their goals and making others succeed through their actions. Confidence and desire to achieve is a very effective motivation, and this positive energy is conveyed to your customers.

Be selective and be clear about which personalities traits you want to establish your world-class sales team. Confidence without arrogance is the formula for the finest wine sales representatives worthy of building a team around.

Typically, an outstanding wine sales representative cannot sell wine to all the differing business scenarios they are thrown into. Many management teams make the mistake of placing a highly talented sales representative in the roles they will not succeed within. They will most likely fail because the situations are very different and require a special skill set. I have experienced very talented sales representatives fail in wine because they were not matched up with their finest abilities. For example, a person selling large volume, low value wines to Costco usually will struggle when trying to sell fine wines to white tablecloth accounts. So, earlier you defined your target sales channels. This is excellent knowledge to have while building a sales team. Place your best people in the areas they can succeed in.

There are a few variations of wines sales roles, usually being defined by who they will be selling to. Here is a high-level outline of the majority of wine sales roles, the types of wine companies that would be an ideal match, who they would be selling to, and the best sales personality and characteristics needed to be successful in each.

1. *National Account Representative*
   These representatives sell to the largest accounts in the country that have national coverage. These wine companies will split on and off premise, but the majority have one representative call on both.

   Top Characteristics:
   Considered a very senior position for only the most experienced reps.
   - Thinks more long-term and strategically than tactically.
   - Is very patient with allowing the process to unfold and pivoting when needed.
   - Enjoys getting into more of the psychology of the deal as well as understanding the people involved.
   - Can psychologically appreciate the larger, less-frequent wins than having more immediate and frequent gratification.
   - High-risk personality.
   - Less passionate and knowledgeable about the wine and more about winning the sale.
   - Loves large volume sales wins (and who doesn't.)

- Enjoys, or can deal with, lots of travel, funky hotels, people coughing on airplanes, and traffic getting home from the airport.

***Ideal Wine Company Matches*:**
Large or mid-sized Wineries, Négociants, Wholesalers or Importers with goals of selling huge volumes of cases at the highest levels.

2. ***Regional / Specialty Account Representative***
Selling to the largest regional or specialty accounts that have significant coverage but are more localized across a few states. Usually on and off premise roles are under one representative's management and responsibility. If a regional account, the representative has a single or contiguous, multi-state patch of grass to manage, or if a specialty account, the representative could have a national reach with the buying decision in one location usually within their region.

***Top Characteristics*:**
Considered a senior to mid-level position for reps with a few years or longer experience.
- Thinks more semi-long-term and strategically than tactically.
- Can psychologically appreciate the larger, less-frequent wins than having more immediate and frequent gratification.
- Semi-higher risk personality.
- Likes large account strategic sales, but more frequent than a few times per year.
- Not as much into the wine, and more into the "wine deal".
- Loves large to medium-volume sales wins.
- Likes to travel, but short state-by-state flights.

***Ideal Wine Company Matches*:**
Large or mid-sized Wineries, Négociants, Wholesalers or Importers with goals of selling significant volumes of cases at the state and regional levels.

3. ***Luxury / Prestige Account Representative***

Very specialized role for only the most knowledgeable, cork-dorkiest, and possibly the snobbiest of wine savvy sales representatives. This requires them to have a very high wine IQ and the ability to wax poetic about details regarding wine, wine regions, wine people, wine whatever.

Top Characteristics:
Considered a senior to mid-level position for very experienced, highly wine knowledgeable reps.
- Very passionate and knowledgeable about the wine.
- Needs to love the product their selling.
- Patience and resilience are important, but the quality of the sales matters more.
- Has a comprehensive fine wine knowledge above most, particularly in the wine business.
- Can deal with large, and I mean very large, buyer egos because they know they will be selling to the most knowledgeable of people.
- Doesn't mind managing and dealing with strong weekly or monthly sales goals.

Ideal Wine Company Matches:
Very common role for smaller to mid-sized prestige Wineries and Importers. Some négociants play at this level, but not many.

4.   *Broadmarket Account Representative*
This is the daily grind level, but necessary for all newcomers to the business. This is where you earn your chops and learn the markets from the street level. It's not as rough as it seems because there are many wine unsophisticated mom-and-pop businesses that need great wine consultation. Most new representatives start here, and it's seen as a career validating start to many future wine executives.

Top Characteristics:
Considered a mid-level to junior/entry-level position for very new reps looking to grow in wine sales.
- Thinks short-term and is more tactical in nature.
- Loves daily sales goals.

- Wants and needs the immediate sale with constant confirmation.
- Less patient and is able to handle a large amount of activity and information.
- Not as much into the wine, and more into the "wine deal".
- Usually a great place to start in wine sales. Many more open career options in this market.

***Ideal Wine Company Matches*:**
Large to mid-size Wholesalers usually have the largest teams with the most comprehensive coverage. Very large Wineries, depending upon their go-to-market strategies, will sometimes have large Broadmarket teams but usually leave it up to their Wholesale partners in each state. Importers, large and small, sometimes have Broadmarket teams as well but usually in the states they are headquartered.

5. ***D2C Sales Representative***
This is a catchall category that, with the recent growth of online sales, can take on a few different forms. The most obvious is a winery tasting room or retail, taking care of the visitors and club members. The newest and most innovative is for a winery or business that sells online. Though they are not seeing customers face-to-face and it skews to a more marketing type role, they still interact and sell wine with important buyers.

***Top Characteristics*:**
Considered a mid-level to junior/entry-level position for very new reps looking to grow in wine sales or has other means of income and can only work part-time.
- Has a strong intuitive sense of the wine consumer.
- Loves interacting with large groups of diverse and sometimes drunk, annoying visitors, fielding random wine questions.
- Has a high sense of and enjoys online marketing.
- Can deal with a daily sales goals and sometimes frantic paces.
- Doesn't mind working in a structured environment, albeit sometimes in crazy beautiful places like Napa Valley or in a NY downtown SoHo hipster office with other too cool for the room people.

***Ideal Wine Company Matches*:**
All Wineries or Négociants with tasting rooms, Retailers and all D2C wine companies.

6. ***Jack-or-Jill-Of-All-Trades Representative***
The final career path, and one that is one of the most common amongst smaller to mid-sized wine businesses, is the "please handle all of our retail and wholesale sales" representative. Wine companies love to task smart, passionate people with way too much to do, and this role is the perfect example. It's usually a lower-paying but highly regarded, learn all aspect of the wine business type of role . . . which, if this is what you're looking for, and can handle the lower income for a period of time, is usually a great career option.

***Top Characteristics*:**
Considered a mid-level to junior/entry-level position for very new reps looking to grow in wine sales or has other means of income and can only work part-time.
- Can handle a significant amount of activities and balls in the air.
- Wants to learn about many parts of the wine business, not just sales.
- Thinks short-term and is more tactical in nature.
- Wants and needs the immediate sale with constant confirmation.
- Needs to be very patient and okay with the fact he/she will not get to everything that needs to be done.
- Not as much into the wine, and more into the "wine deal".
- Usually a great place to start in wine sales. Many more open career options in this market.

***Ideal Wine Company Matches*:**
Small to mid-sized Wineries, Wholesalers, D2C and Importers that are spread fairly thin.

There is someone for everyone. Choose wisely, because being a good wine sales representative is good, but having the right people in the right roles is everything. Also, finding sales representatives that are accountable to the expectations is very difficult. What that means is a salesperson who cares, is smart, passionate, confident and not scared or intimated to take on weekly or monthly goals, works very hard towards them, then stands tall and accepts their strengths and successes as well as failures. With wine sales comes many expectations, sometimes unreasonable. They have the ability to take in the good with the tough and the bad, focus on goals, hold themselves accountable, accept results, have a bit of amnesia and work towards the next goal. These are some of the finest characteristics that distinguish a so-so representative from an exceptional one.

Find sales professionals that are full of passion and constantly seeking perfection. And, most importantly, match what the internal role requires to the experience a representative possesses. It is imperative to their, and your, overall success.

3 Imperatives Of Embracing Account Target Lists

Again, sales representatives are hunters by nature. They have the personality and drive to go after what they need to be successful. They are constantly in motion and driven by the need to achieve. Those are great qualities to possess. But if you don't point them in the right direction they will still hunt, just not as effectively.

It requires you to develop a culture of respecting and supporting a sales-driven business that will enable the team to work in concert and empower them to know that they can achieve even greater heights together. It is not an easy team to wrangle, but once they are in a groove, they will return with riches and stories of conquering the industry that will lift all in your business. Nothing is more satisfying to the team than hearing your wine is making it to and being enjoyed by customers. It's also nice for them to know that everyone will be getting their paychecks for the foreseeable future. Happy team, happy customer, happy wine, happy bottom line.

So, there are two essentials to get the most from some of your highest paid employees, your sales team.

- *Having well-defined and detailed target lists*
  Defined, meaning down to the name of the accounts in every marketplace. They should have a list of their top 50, 100, 300, whatever the number is based upon the strategy, of the accounts they are targeting each and every day. I've seen wine companies loosely define it as a higher category, like go after "luxury restaurants", only to find it difficult for the sales representatives to focus, and impossible for management to measure. In addition, the wholesale representatives like these lists because it makes their goals very clear and actionable. With your account targets very well-defined, let the hunt begin, and you will find the more successful sales representatives will embrace this exercise and reveal themselves quickly.

- *Prospecting is an all-the-time thing*
  Without getting too far into sales philosophy, instilling a culture that believes prospecting to be the most important and time-consuming sales activity of your team is incredibly important. It's the most challenging duty, even for the best sales representatives, and I've only known a few who actually enjoy doing it. However, without outstanding and ongoing prospecting, the whole model breaks down. It sounds fairly straightforward, but studies have shown that, other than presenting in front of an intimidating buyer, it can be the scariest activity a sales representative must perform. And you're asking them to do it every day. Be understanding and patient but be firm that sales prospecting is a pursuit that happens every day. And if you have software to measure it, you'll be way ahead in the wine sales game.

Having a very disciplined sales culture that is focused on prospecting is truly imperative and will be the most important exercise your team performs regularly.

When I worked for larger wine companies, we were tasked with categorizing retailers down to very specific types of businesses, like "white tablecloth luxury restaurants". Then the brand managers would put together a sales strategy for each type. It was very detailed because the brands had wide distribution goals that needed to be dictated as such. I fully understand these were most likely much different business structures with considerably more resources than yours, and I do not believe in planning to this much detail because it takes away from other more important tasks needing attention. However, I think some of the disciplines I learned in the larger wine companies apply to all, regardless of size or ambitions. Nothing will offer you an opportunity for greater success than this single initiative. Having the team create and manage a list of your top prospects could be one of the greatest disciplines you indoctrinate into the culture of your sales team. Define this for everyone so there are no misconceptions. Most importantly, make this measurable for the sales and management teams. Everyone is invested and incented to sell. Don't leave anything to chance. There are enough risks and unknowns in the wine business. It's very empowering to hold a complete list of the retail accounts your wine needs to be sold and poured in. It's enlightening, motivational, actionable and measurable.

There are three simple steps toward creating, managing, and driving progress through the use of target accounts lists.

1. ***Work with wholesalers to create an account target list per market***

    Have your sales teamwork with each of your wholesale partners to create a simple top 25-50 target account list. This should be by state, so you will have a target list for each marketplace. The list should be comprised of accounts that match your brand positioning, strategy, and pricing expectations. A good way to know if it is a good prospective account is to determine if other similar brands excel in these places. If you're selling a $100 Napa Cabernet Sauvignon and the account has an excellent reputation of selling higher end Napa Cab's, they should be on your account list. Also, the list should be very simple and, ideally, on one-page. The details needed are your representative's name, wholesaler representative's name, account name, buyer name, last time contacted, method of contact, meeting date and closed date. It should fit on one line easily.

2. ***Use the list to drive daily, weekly and monthly activities***
This list should be the primary playbook used for almost all of your teams' sales activities. Nothing should matter more, and rightfully so. As this list is managed and accounts are closed, one by one, wine is being sold and goals are being achieved. It is putting your companies targets into real, tangible sales activities. In weekly sales meetings, have your team use the lists as their primary source of information, updating the team on each accounts progress. Some of these accounts your sales representative will be selling to directly, and other accounts the wholesale representatives will be selling to. Your team should be highly knowledgeable and on top of each and every prospect.

Referring back, this will require your team to work with your wholesale partners at least once per month to update the list and hold them accountable for very focused progress.

3. ***Motivate the team by incenting target account list progress***
Nothing puts a red spotlight on the importance and urgency of a sales initiative like dollars. Depending upon your financial situation, put very specific non-revenue goals in place for your team to achieve as it directly pertains to making progress on their lists. For example, each month have $1,000 commission available to each of your sales representatives. To achieve this, they have to schedule and conduct 20 customer meetings, close 5 new accounts, and have their wholesale representative close 5 new accounts. Use whatever specific goals are right for your situation, but as you can see, using this method creates a very factual and measured way to incent and manage your team, as well as dictate your own success.

To emphasis the importance of this exercise, this is the most effective sales methodology and tool I have ever experienced. It is tangible, actionable, motivational and creates real sales dollars when done well.

As an important side point, regardless of the size of your wine company, I strongly suggest getting a great customer relations and sales management (CRM) software package for your team to use. There are plenty of online-based solutions available. This will give you all the tools needed to create the strategy and make your team very successful. You will be able to delve deep inside the sales side of your business for planning and performance analysis and have a solid history that new hires can jump into in the event of turn-over, which is more common in sales than most other parts of the business. Finally, it allows you to stay out of the weeds of the sales process, letting you focus on the results of the activities and successes.

FINAL THOUGHTS

"Whatever you do, pour yourself into it."
- Robert Mondavi

So, there you have it - a rare occurrence in the world of wine. A master class on sales and marketing. To most in the wine business, I believe they understand the importance of this subject matter but it's not the sexiest of topics in an industry known so much more for the artistry and amazement of the wine. This subject of sales and marketing I chose to write about is the behind-the-scenes undertakings that few really consider early on… *until it's too late.*

But I believe that if you made it this far, there is a reason for you to be engaged. Maybe you have a wine brand that is struggling, or you're considering a new wine venture that you dream will be renowned around the world for its luxury wine and experience. Whatever the purpose, you have taken what I believe to be the most important step toward your success by diving deeper into this very essential topic. It's kind of a big statement to make when considering the huge amount of time, effort and expense required to put together a wine, great brand and state-of-the-art winery. However, I'm very comfortable making that statement because I've seen too many great efforts fail for nothing more than a lack of understanding and passion about developing a strong sales and marketing strategy early on.

I set out to write this guide for a few reasons.

Firstly, I've been fortunate to gain all of this knowledge over the many years in this amazing industry, and if I was able to help you in any way, then to me it was a very worthy effort. Thank you for allowing me to share this with you and "pay it forward" to an industry that I adore.

Secondly, I did it for the love of wine.

It's a remarkable and humbling thought to me that I was fortunate to find what captured my heart, my sole and my mind all in one. I've heard it doesn't happen very often in many people's lives, but for some divine reason, it did for me. I really don't know why. Maybe it was completely random or maybe destiny does exist. I may never really know, which is okay to me because I'm here now, living this incredible life each day completely engaged and in love with what I do for a career, a hobby and a larger purpose.

It's clear you have been fortunate as well. So, finish your goblet of wine and let's go sell some wine together while we live the dream!

Cheers - Eric

ABOUT THE AUTHOR

ERIC GUERRA *(aka: Galactic Viceroy Of Wine)*

Named the 2018 North Bay Business Journal Winery Sales and Marketing Officer Of The Year, with over 17 years of wine experience having managed or sold over 20 million cases and developed over 500+ wine brands . . . author, executive, marketing and sales professional, wine blogger, wine author, wine educator, fine wine investor, and wine public speaker Eric Guerra has led some of the industries' most iconic wineries and wine brands.

His roles have been diverse, from three-tier and D2C sales, brand marketing, tasting room, finance, sourcing, winemaking to production.

Eric's most recent venture is the co-developer and partner of the Reserve Tastings Wine Company LLC, a successful direct-to-consumer private Wine + Music club that offers high-end wines that are paired with musically inspired label artwork and playlists.

In his free time Eric . . . drinks wine.

"Sip Generously, Listen Loudly… and Repeat Often"
www.reservetastings.com

www.ingramcontent.com/pod-product-compliance
Lightning Source LLC
Chambersburg PA
CBHW021712210326
41599CB00013B/1629